SRA Algebra Reddiness

Assessment

Sharon Griffin

SRA

Columbus, OH

Author
Sharon Griffin
Professor of Education and
Adjunct Associate Professor of Psychology
Clark University
Worcester, Massachusetts

Mathematics Content Standards for California Public Schools reproduced by permission,
California Department of Education, CDE Press, 1430 N Street, Suite 3207, Sacramento, CA 95814.

SRAonline.com

Printed in the United States of America.

Send all inquiries to this address:
SRA/McGraw-Hill
4400 Easton Commons
Columbus, OH 43219

ISBN: 978-0-07-612397-1
MHID: 0-07-612397-9

1 2 3 4 5 6 7 8 9 DBH 12 11 10 09 08 07

The McGraw·Hill Companies

Contents

Algebra Readiness
Assessment Overview

Algebra Readiness is rich in opportunities and resources to conduct comprehensive assessments that inform instruction.

Goals of Assessment

1 To improve instruction by informing teachers about the effectiveness of their lessons

2 To promote growth of students by identifying where additional instruction and support are needed

3 To recognize accomplishments

Algebra Readiness assessments have the following characteristics to ensure reliable feedback:

- **Rich Variety** Assessments cover five math proficiencies—understanding, reasoning, computing, applying, and engaging—so teachers can assess a rich variety of mathematics topics.

- **Multiple Sources** Numerous assessment types are included in the program so teachers can gather evidence from many sources.

- **Alignment** Assessments are carefully aligned with the curriculum and accurately test the program material. The curriculum is correlated to national and state standards so the assessments include the required content.

Phases of Assessment

Plan While developing lesson plans, you can determine how to assess students' grasp of the lesson material.

Gather Evidence Throughout the instructional phase, you can gather evidence of student comprehension. The end of every lesson is designed to help you conduct meaningful assessments. The Informal Assessment checklists and Student Assessment Records are provided to help you accurately record data.

Summarize Findings Taking time to reflect on the assessment results to plan follow-up activities is a critical element of any lesson.

Use Results Use the results of your findings to differentiate instruction or to adjust or confirm future lessons.

Number Worlds Assessments

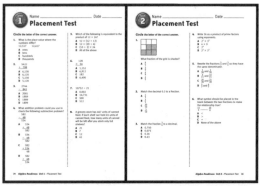

Placement Test, Algebra Readiness

- **Placement Tests** identify whether or not students should begin the *Algebra Readiness* curriculum.

- **Weekly Tests** assess student comprehension of the week's five daily lessons.

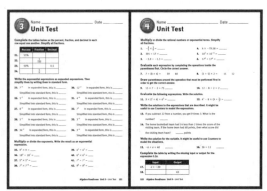

Unit Test

- **Diagnostic Tests** in every lesson assess progress for immediate differentiation.

- **Unit Tests** evaluate concept acquisition for the entire unit. These tests are available in both open-response and multiple–choice formats. They measure cumulative progress throughout the year.

- **Rubrics,** shown on pages 9–11, are available in every lesson to informally evaluate student understanding.

Using the *Number Worlds* Placement Tests

Each level of **Number Worlds** contains a placement test to determine whether students understand the content in each level of the program. Every placement test assesses knowledge of two adjacent levels of **Number Worlds** so you can observe and compare information about which level's content is appropriate for each student.

Placement Tests in Levels A–C

Number Worlds Levels A–C are targeted for use by students in Grades PreKindergarten through Grade 1. Placement of students at these levels should begin by evaluating their success on the Placement Tests on pages 12–23.

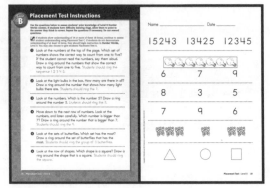

Placement Test, Level B

In Levels A–C, the placement tests are designed to be administered individually to children by a teacher, classroom aide, or parent helper. The tests at these levels consist of teacher's instructions and reproducible test masters. The test questions for these levels are intended to be read to the students while the test administrators track students' answers on the Placement Test Record on page 99.

If students demonstrate understanding of more than 75% (at least 14 of 18) of the test items, continue to assess student comprehension using the test for the next level. If students do not demonstrate understanding of at least 14 items, they should begin instruction in that level of the **Number Worlds** Program.

To gain an even more thorough understanding of a student's knowledge of number, use the **Number Knowledge Test.**

The **Number Knowledge Test** was designed to assess central conceptual knowledge typically acquired by children around the ages of 4, 6, 8, and 10 years.

Placement Tests in Levels D–J

Number Worlds Levels D–J are targeted for use by students in Grades 2–8. Placement of students at these levels should begin by evaluating their success on the reproducible Placement Tests on pages 12–23.

You may choose to assess students on only one topic of concern or assess students on their knowledge of several topics at each level.

In Levels D–J, the placement tests may be administered to children by a teacher, classroom aide, or parent helper. The assessment consists of topic-specific tests at two levels. The first page assesses knowledge of content in the previous level, and the second page assesses knowledge of the content covered at this level.

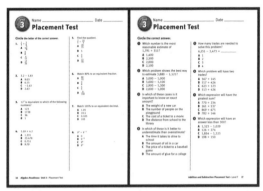

Placement Test, Level F

Placement Test in Algebra Readiness

Algebra Readiness is targeted for use by students who need basic skills reinforcement in order to be successful in Algebra. Placement of a student into *Algebra Readiness* or an Algebra I program should begin by evaluating student success on the reproducible Placement Tests on pages 14–19.

Students who do not successfully complete 75% of the items on the test should be placed into the *Algebra Readiness* program. If students successfully complete 75% of the items, you may choose to have them begin in the Algebra Readiness program or place them into Algebra I.

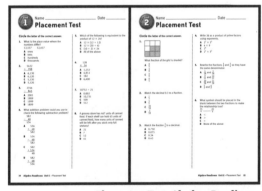

Placement Test, Algebra Readiness

Formal Assessments

Algebra Readiness provides formal assessments for each chapter.

- **Chapter Tests** evaluate students' understanding of chapter concepts.

- **Unit Tests** are available to assess students' understanding of several chapters.

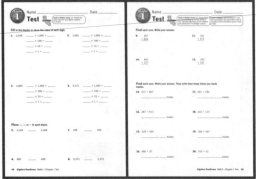

Chapter Test, Algebra Readiness

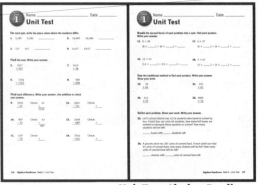

Unit Test, Algebra Readiness

Chapter Tests

Chapter Tests assess students' knowledge of the lesson content for each week. These tests help identify where students are still having difficulty so additional quick intervention can effectively enhance performance.

Unit Tests

Unit Tests assess students' knowledge of the topics in each level. If students are still having trouble with a concept, these tests identify the lesson in which remediation is available.

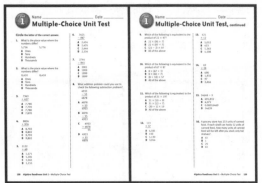

Multiple-Choice Unit Test, Algebra Readiness

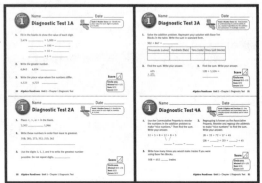

Diagnostic Tests, Algebra Readiness

Multiple-Choice Unit Tests

Multiple-Choice Unit Tests may be used in place of Unit Tests or as an additional assessment after remediation. Multiple-Choice Unit Tests are also useful for providing practice with standardized-format assessment.

Diagnostic Tests

Use **Diagnostic Tests** in every lesson assess progress for immediate differentiation.

Informal Daily Assessment

Through day-to-day observations, Informal Daily Assessments evaluate how well each student is learning and grasping skills. The Engage activities, for example, allow you to watch students practice particular skills under conditions that are more natural than most classroom activities. Warm–Up exercises allow you to see individual responses, give immediate feedback, and involve the entire class.

Simple rubric checklists enable teachers to record and track observations. These can be recorded on the Student Assessment Record on page 100 to help provide a more complete view of student proficiency. By the end of each chapter, teachers who use these assessments will have had several opportunities to record evidence of student achievement for every student in all five math proficiencies.

The following are general analytic rubrics for different math proficiencies. These rubrics are provided to help teachers identify key behaviors that indicate levels of proficiency with the intention of differentiating instruction.

PROFICIENCY	ACTIVITY	RUBRIC CHECKLIST
Computing	Warm Up Engage	Responds accurately Responds quickly Responds with confidence Self-corrects
Understanding	Engage	Makes important observations Extends or generalizes learning Provides insightful answers Poses insightful questions
Engaging	Engage	Pays attention to the contributions of others Contributes information and ideas Improves on a strategy Reflects on and checks accuracy of work
Reasoning	Engage Reflect	Provides a clear explanation Communicates reasons and strategies Chooses appropriate strategies Argues logically
Applying	Engage Reflect	Applies learning to new situations Contributes concepts Contributes answers Connects mathematics to the real world

Computing Rubric

CRITERIA	4	3	2	1
Computational Accuracy	Correct all of the time	Correct most of the time	Correct some of the time	Mostly incorrect
Computational Speed	Quick response	Often quick	Slow response	Inappropriate or incorrect
Flexibility	Thinks of different ways to solve a computation	Thinks of one other way to solve a computation	Focuses on only one way to solve a computation	Has attempted to memorize a way to solve a computation
Appropriateness	Chooses an appropriate operation to solve a computation	Mostly chooses an appropriate operation to solve a computation	Sometimes chooses an appropriate operation to solve a computation	Rarely chooses an appropriate operation to solve a computation

Engaging Rubric
Math Class Participation

CRITERIA	4	3	2	1
Promptness in Math Assignments	Always prompt in completing assignments	Occasionally late returning work assignments	Frequently late handing in work assignments	Rarely completes work
Preparation for Math Class	Almost always prepared for class with assignments and required materials	Usually prepared for class	Rarely prepared for class	Consistently unprepared for class
Behavior during Math Class	Almost never displays disruptive behavior during class	Rarely displays disruptive behavior	Occasionally displays disruptive behavior	Constantly displays disruptive behavior
Level of Engagement in Mathematics	Contributes to class and guided discussion by offering ideas and asking questions on a regular basis	Occasionally contributes by offering ideas and asking questions	Rarely offers ideas or asks questions	Never contributes ideas or asks questions

Understanding Rubric

Communicating Mathematically

CRITERIA	4	3	2	1
Explanations	Rich, precise, and clear	Accurate and mostly clear	Appropriate occasionally, but may not be clear	Unclear or inappropriate
Representations	Very perceptive using charts, diagrams, or graphs	Accurate and appropriate	Appropriate but imprecise	Inappropriate or incorrect
Conclusions	Logical and appropriate	Mostly clear and logical	Some clear and understandable conclusions	Conclusions are not clear or are inappropriate
Math Language	Uses the language of mathematics to explain concepts	Uses some math terms in explanations	Includes few math terms in explanations	Does not use the language of mathematics

Reasoning Rubric
Problem Solving

CRITERIA	4	3	2	1
Understands Problems	Identifies real problems independently	Independently recognizes there is a problem	Waits for someone else to recognize a problem	Cannot identify a problem even when someone else does
Gathers Facts	Accesses information to obtain all necessary facts	Knows where to look to obtain facts	Able to gather one or two facts independently	Does not realize the need to find facts and uses any set of numbers to calculate
Brainstorms Solutions	Generates several creative solutions independently	Generates a solution independently	Generates ideas with assistance	Does not generate any solution and may ask, "What do I do?"
Problem Solving	Develops an efficient and workable strategy	Develops a workable strategy	Uses an appropriate strategy some of the time	Uses an inappropriate or unworkable strategy
Evaluates Solutions	Uses reflection to adjust methods	Takes time to analyze effectiveness of each possible solution	Recognizes pluses and minuses of some solutions with assistance	Does not evaluate the effectiveness of proposed solutions
Persistence	Continues problem solving outside of math class	Persists in class independently	Will try with encouragement and assistance	Gives up easily

Applying Rubric

Applications of Mathematics

CRITERIA	4	3	2	1
Real World	Identifies ways in which math is used outside of math class	Identifies ways in which math is applied in the real world when asked	Recognizes that math is applied when directed	Does not see relevance of math to real-world problem solving
Integration of Mathematics	Makes connections among the different strands of mathematics	Identifies connections among strands of mathematics when asked	Understands connections among mathematics when directed	Cannot understand connections among strands of mathematics
Problem Solving	Uses math as a key strategy to solve real-world problems	Uses math with assistance to solve real-world problems	Understands that math can be used to solve real-world problems	Cannot understand how mathematics is useful in real-world problem solving

Name _____   Date _____

Placement Test

Circle the letter of the correct answer.

1. What is the place value where the numbers differ?

12,517   12,417

A ones
B tens
C hundreds
D thousands

2.   5422
      + 708

A 6,130
B 6,120
C 5,130
D 5,120

3.   2744
      − 845

A 2001
B 1909
C 1999
D 1899

4. What addition problem could you use to check the following subtraction problem?

      582
    − 48
      534

A   534
    + 48
      582

B   534
    − 48
      582

C   582
    + 534
       48

D   582
    + 48
      534

5. Which of the following is equivalent to the product of 12×24?

A $12 \times (12 + 12)$
B $12 \times (20 + 4)$
C $(10 + 2) \times 24$
D All of the above

6.   128
      × 54

A 1,152
B 6,912
C 182
D 6,400

7. $10752 \div 21$

A 0.002
B 10,731
C 500
D 512

8. A grocery store has 467 units of canned food. If each shelf can hold 65 units of canned food, how many units of canned will be left after you stock only full shelves?

A 21
B 7
C 12
D 41

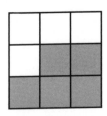

Name _____ Date _____

Placement Test

Circle the letter of the correct answer.

1.

What fraction of the grid is shaded?

A $\frac{9}{5}$

B $\frac{5}{9}$

C $\frac{4}{9}$

D $\frac{4}{5}$

2. Match the decimal 0.3 to a fraction.

A $\frac{1}{3}$

B $\frac{3}{5}$

C $\frac{30}{10}$

D $\frac{3}{10}$

3. Match the fraction $\frac{3}{4}$ to a decimal.

A 0.750

B 0.075

C 0.34

D 0.43

4. Write 36 as a product of prime factors using exponents.

A $2^2 \times 3^3$

B 4×9

C 2^4

D $2^2 \times 3^2$

5. Rewrite the fractions $\frac{5}{7}$ and $\frac{3}{4}$ so they have the same denominator.

A $\frac{5}{28}$ and $\frac{3}{28}$

B $\frac{9}{11}$ and $\frac{10}{11}$

C $\frac{20}{28}$ and $\frac{21}{28}$

D $\frac{5}{7}$ and $\frac{6}{7}$

6. What symbol should be placed in the blank between the two fractions to make the relationship true?

$\frac{15}{4}$ _____ $\frac{20}{6}$

A $<$

B $>$

C $=$

D None of the above

Name _____ Date _____

Placement Test

Circle the letter of the correct answer.

1. $\frac{1}{3} + \frac{3}{8}$

   A $1\frac{3}{8}$

   B $\frac{4}{8}$

   C $\frac{4}{11}$

   D $\frac{17}{24}$

2. $3.2 - 5.83$

   A 9.03

   B 6.15

   C -2.63

   D 2.63

3. 12^3 is equivalent to which of the following numbers?

   A 123

   B 1728

   C 36

   D 1

4. 5.59×4.2

   A 1.331

   B 23.478

   C 0.751

   D 9.79

5. Find the quotient.

   $\frac{2}{5} \div \frac{10}{3}$

   A $\frac{20}{15}$

   B $\frac{6}{50}$

   C $\frac{4}{3}$

   D $\frac{15}{20}$

6. Match 80% to an equivalent fraction.

   A $\frac{80}{10}$

   B $\frac{0}{8}$

   C $\frac{4}{5}$

   D $\frac{8}{100}$

7. Match 105% to an equivalent decimal.

   A 1.05

   B 10.5

   C 0.105

   D 105

8. $9^3 \div 9^{-3}$

   A 0

   B 9^{-9}

   C 9^6

   D 9^3

Name _____ Date _____

Placement Test

Circle the letter of the correct answer.

1. Evaluate the expression by completing the operations in the correct order.
   $(10 + 3) \times 2$
   A 16
   B 26
   C 15
   D 23

2. What should the parentheses be placed around to get the correct answer?
   $6 - 12 \div 3 = 2$
   A $-$
   B $\div$
   C $6 - 12$
   D $12 \div 3$

3. $3 \times (4 + 5) + 2^3$
   A 24,389
   B 25
   C 35
   D 6,859

4. If you add 15 to a number, you get 5 times 12. What is the number?
   A 75
   B 240
   C 45
   D 185

5. Solve for the variable.
   $-7a = 182$
   A 26
   B -26
   C 189
   D 175

6. What is the input for the expression $x \times \frac{1}{3}$ if the output is 10?
   A 3
   B 30
   C 0
   D $\frac{3}{10}$

7. Name the property that is used to change Expression 1 into Expression 2.
   Expression 1: $\frac{1}{8} \times 8$
   Expression 2: 1
   A identity property of multiplication
   B commutative property of multiplication
   C inverse property of multiplication
   D associative property of multiplication

8. Solve for the variable.
   $6x - 6 = -24$
   A 3
   B -5
   C -3
   D 5

Placement Test

Circle the letter of the correct answer.

1. How do you move from (5, 3) to (−4, 10)?

    A Left 9 units and up 7 units

    B Right 9 units and down 7 units

    C Right 7 unit and up 9 units

    D Left 1 unit and up 7 units

2. What point is graphed on the following coordinate grid?

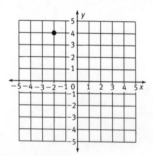

    A (−2, 4)

    B (4, 2)

    C (2, 4)

    D (4, −2)

3. Which of the following points could be connected to create a square?

    A (1, 8), (8, 1), (4, 5), (5, 4)

    B (1, 3), (4, 6), (1, 6), (4, 3)

    C (9, 10), (3, 5), (9, 5), (3, 10)

    D (1, 1), (2, 2), (3, 3), (4, 4)

4. Determine the length of the line segment that connects (5, −2) and (−5, −2)

    A 0

    B −10

    C 10

    D 4

5. Find the length of the hypotenuse if the lengths of the other two sides are 10 and 24.

    A 26

    B 14

    C 34

    D 12

6. Which of the following is a graph of $y = -3x + 2$?

    A

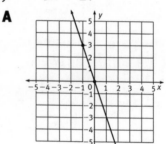

    B

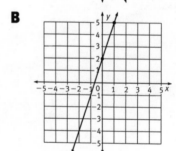

    C

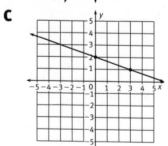

    D

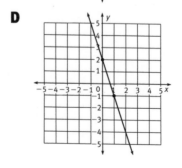

Name _____ Date _____

Placement Test

Circle the letter of the correct answer.

1. Match $\frac{1}{7}$ with its opposite.

   A 7

   B $-\frac{1}{7}$

   C -7

   D $\frac{1}{7}$

2. Match $\frac{1}{7}$ with its reciprocal.

   A 7

   B $-\frac{1}{7}$

   C -7

   D $\frac{1}{7}$

3. Simplify the variable expression.

   $$\sqrt{100x^4y^6}$$

   A $100xy$

   B $10x^2y^4$

   C $10xy$

   D $10x^2y^3$

4. Find the missing side of the right triangle. The hypotenuse is c.

   $a = 5$; $c = 9$; find b.

   A 4

   B $\sqrt{56}$

   C $\sqrt{106}$

   D 14

5. Simplify. Assume all variables do not equal zero.

   $-z^3(z^5x^4 + z^2)$

   A $z^2x^4 + z$

   B $-z^8x^4 - z^5$

   C $z^8x^4 + z^5$

   D $-z^{15}x^4 - z^6$

Solve for the variable.

6. $12x + 3x - 8x = x + 30$

   A $\frac{30}{7}$

   B -5

   C 5

   D 0

7. $9(x + 3) = 0$

   A 3

   B 9

   C 0

   D -3

8. Solve for the inequality.

   $-(y + 5) - 12 < -2$

   A $y > -15$

   B $y < 15$

   C $y = 15$

   D $y > 15$

Chapter 1
Diagnostic Test 1A

Name _____ Date _____

1. Fill in the blanks to show the value of each digit.

   2,476 _____ × 1,000 = _____

   _____ × 100 = _____

   _____ × 10 = _____

   _____ × 1 = _____

2. Write the greater number.

   6,843   6,834 _____

3. Write the place value where the numbers differ.

   4,523   4,723 _____

· ·

Chapter 1
Diagnostic Test 2A

Name _____ Date _____

1. Place <, >, or = in the blank.

   2,302 _____ 1,966

2. Write these numbers in order from least to greatest.

   358; 281; 272; 352; 253; 262

   _____

3. Use the digits 3, 5, 2, and 9 to write the greatest number

   possible. Do not repeat digits. _____

Chapter 1

Name ______________________ Date ____________

Diagnostic Test 3A

1. Solve the addition problem. Represent your solution with Base-Ten Blocks in the table. Write the sum in standard form.

302 + 847 = __________

Thousands (cubes)	Hundreds (flats)	Tens (rods)	Ones (unit blocks)

2. Find the sum. Write your answer.

$$\begin{array}{r} 435 \\ +\ 271 \\ \hline \end{array}$$

3. Find the sum. Write your answer.

128 + 1,524 =

____

Score /3

Circle one.
Minimal (0/3–1/3)
Basic (2/3)
Secure (3/3)

Chapter 1

Name ______________________ Date ____________

Diagnostic Test 4A

1. Use the Commutative Property to reorder the numbers in the addition problem to make "nice numbers." Then find the sum. Write your answer.

12 + 5 + 8 = 12 + 8 + 5

\ /

______ + ______ = ______

2. Regrouping is known as the Associative Property. Reorder and regroup the addends to make "nice numbers" to find the sum. Write your answer.

28 + 33 + 72 + 17 + 45

(28 + ______) + (33 + ______) + 45

______ + ______ + ______ = ______

3. Write how many times you would make trades if you were using Base-Ten Blocks.

508 + 652 ______ trades

Score /3

Circle one.
Minimal (0/3–1/3)
Basic (2/3)
Secure (3/3)

Name _____ Date _____

Diagnostic Test 1A

1. Write the number in standard form.

8,000 + 500 + 50 + 4 _____

2. Use the digits given to make the largest number possible using each digit one time. Write your answer.

2, 1, 8, 0 _____

3. Use the digits given to make the smallest number possible using each digit one time. Write your answer.

8, 6, 1, 4 _____

Score ☐/3

Circle one.
Minimal (0/3–1/3)
Basic (2/3)
Secure (3/3)

- -

Chapter 2

Name _____ Date _____

Diagnostic Test 2A

1. Use Base-Ten Blocks to help you subtract. Draw a model of the problem using Base-Ten Blocks, and cross out the blocks that you are subtracting.

205
− 170

2. Choose the two numbers from the group that have the greatest difference. Write and solve that subtraction number sentence.

407   289   352

_____

3. Choose the two numbers from the group that have the smallest difference. Write that subtraction number sentence.

482   517   308

_____

Score ☐/3

Circle one.
Minimal (0/3–1/3)
Basic (2/3)
Secure (3/3)

Chapter 2

Name ______________________ Date ____________

Diagnostic Test 3A

1. What is an addition problem that you can use to check the following subtraction problem? Write the addition problem.

```
  631
- 141
  490
```

____

2. Find the difference. Write your answer.

```
  278
-  69
```

3. Is the following subtraction problem correct? Write how you checked the answer and whether it is correct.

```
  308
- 123
  175
```

This is ______________________

because ______________________.

Score ___/3

Circle one.
Minimal (0/3–1/3)
Basic (2/3)
Secure (3/3)

Chapter 2

Name ______________________ Date ____________

Diagnostic Test 4A

1. Write the missing digits in the subtraction problem.

```
  284
- 2_2
   22
```

2. Write why you can or cannot solve for the missing digits.

```
  46_
- 3__
  _33
```

____

____

____

3. Write the missing digits in the subtraction problem.

```
  26_
- 1_5
  _53
```

Score ___/3

Circle one.
Minimal (0/3–1/3)
Basic (2/3)
Secure (3/3)

Name _____ Date _____

Diagnostic Test 1A

1. Find the product. Write your product as repeated addition.

  4 × 3

  _____

2. Write how many groups are shown in the array.

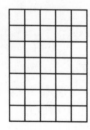

  _____ groups of _____

3. Write the multiplication sentence modeled by the array.

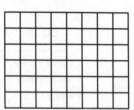

  _____ × _____ = _____

Score ◹ /3

Circle one.
Minimal (0/3–1/3)
Basic (2/3)
Secure (3/3)

· ·

Chapter 3

Name _____ Date _____

Diagnostic Test 2A

1. Break the second number of the problem into a sum. One of the addends should be a multiple of ten. Find the product. Write your answer.

  5 × 24

  (5 × _____) + (5 × _____) = _____

2. Fill in the blanks to find the product.

  30 × 62

  (_____ × _____) + (_____ × _____)

  _____   +   _____

  _____

3. Find the product. Write your answer.

$$\begin{array}{r} 113 \\ \times\ 11 \\ \hline \end{array}$$

Score   /3

Circle one.
Minimal (0/3–1/3)
Basic (2/3)
Secure (3/3)

Name _____   Date _____

Diagnostic Test 3A

1. Write the property shown in each equation.

$13 \times (10 \times 7) =$
$(13 \times 10) \times 7$

$12 \times (20 + 4) =$
$(12 \times 20) + (12 \times 4)$

$315 \times 251 =$
251×315

_____   _____   _____

2. Write both of the multiplication equations that are modeled by the following array.

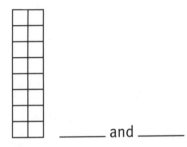

_____ and _____

3. Use the Partial Products Method to find the product. Write your answer. Show your work.

$$\begin{array}{r} 218 \\ \times\ 29 \\ \hline \end{array}$$

Score ◱ /3

Circle one.
Minimal (0/3–1/3)
Basic (2/3)
Secure (3/3)

- -

Name _____   Date _____

Diagnostic Test 4A

1. Use the traditional method to find each product. Write your answer. Show your work.

$$\begin{array}{r} 217 \\ \times\ 22 \\ \hline \end{array}$$

2. Solve the multiplication problem. Write your answer.

$12 \times 23 =$ _____

3. Write a multiplication problem to fill the grid, leaving a few squares empty.

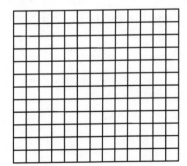

_____

Score ◱ /3

Circle one.
Minimal (0/3–1/3)
Basic (2/3)
Secure (3/3)

Name _____ Date _____

Diagnostic Test 1A

48 people need to be placed into 12 cars for the carpooling groups. How many people should ride in each car so each car contains the same number? Solve the problem by answering each question below. Write your answer.

1. What is the problem asking?

_____

2. Draw circles to model distributing the correct number of people to each car.

3. Write the division sentence modeled by the circles.

_____

Score /3

Circle one.
Minimal (0/3–1/3)
Basic (2/3)
Secure (3/3)

Chapter 4

Name _____ Date _____

Diagnostic Test 2A

There are 41 soccer players trying out for the team. The coach wants to see the players in groups of 9 at most. How many full groups of soccer players can the coach make? Write your answers to the following questions.

1. Is 41 evenly divisible by 9? Explain.

_____

2. Draw circles to model distributing the correct number of people to each group.

3. Write the division sentence modeled by the circles.

_____

Score /3

Circle one.
Minimal (0/3–1/3)
Basic (2/3)
Secure (3/3)

Chapter 4

Name _____ Date _____

Diagnostic Test 3A

1. Use long division to find the quotient. Show your work.
   Write your answer.

   $7\overline{)312}$

2. Solve the problem. Show your work. Write your answer.

   There are 102 people going on a trip to the recycling center. A total of 20 people can be seated in each bus. How many full buses of people will go on the trip? How many people will ride in the last bus?

   _____ full buses _____ people will ride on the last bus

3. Write the division sentence that is described in Question 2.

   _____

Score ⊿/3

Circle one.
Minimal (0/3–1/3)
Basic (2/3)
Secure (3/3)

Chapter 4

Name _____ Date _____

Diagnostic Test 4A

The school bought 30 cases of soup for the local soup kitchens. The cases of soup need to be distributed evenly among 4 soup kitchens.

1. Write and solve the division problem described.

   _____

2. Write how the remainder should be handled in this situation

   _____

3. How many cases of soup should each soup kitchen receive?

   _____ cases of soup per soup kitchen

Score ⊿/3

Circle one.
Minimal (0/3–1/3)
Basic (2/3)
Secure (3/3)

Name _____ Date _____

Diagnostic Test 1A

1. Write the number of parts for the whole or the set.

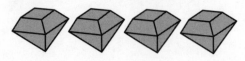

_____ parts

2. Write the fraction that is modeled.

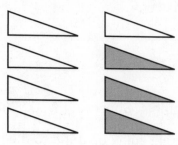

_____

3. Divide the whole set into 6 equal parts by circling individual sets.

Score 3

Circle one.
Minimal (0/3–1/3)
Basic (2/3)
Secure (3/3)

· ·

Name _____ Date _____

Diagnostic Test 2A

1. Draw lines to divide the whole into the given number of equal parts.

6   ◯

2. Draw the fraction as part of a whole.

$\frac{2}{5}$

3. Draw the whole set. Shown is $\frac{1}{6}$ of a set.

Score [   ] 3

Circle one.
Minimal (0/3–1/3)
Basic (2/3)
Secure (3/3)

Name _____ Date _____

Diagnostic Test 3A

1. Use the grid to determine how many fractional parts are needed to equal the whole. Write your answer.

_____ of the grid is shaded.

_____ parts are needed to equal the whole.

_____ is equal to the whole.

3. Draw a grid for the following fraction.

$\frac{9}{9}$

2. Make a whole using the same unit fraction. Write how many more Fraction Bars you will need to make 1.

1

| $\frac{1}{5}$ | $\frac{1}{5}$ | $\frac{1}{5}$ | $\frac{1}{5}$ |

_____ more $\frac{1}{5}$ Fraction Bars

Score ⬜/3

Circle one.
Minimal (0/3–1/3)
Basic (2/3)
Secure (3/3)

Algebra Readiness Unit 2 • Chapter 5 Diagnostic Test   **29**

- -

Name _____ Date _____

Diagnostic Test 4A

1. Write the fraction of a dollar for the given coins.

_____

2. Write the total using cent notation.

_____

3. Write the total using dollar notation.

_____

Score ⬜/3

Circle one.
Minimal (0/3–1/3)
Basic (2/3)
Secure (3/3)

Algebra Readiness Unit 2 • Chapter 5 Diagnostic Test   **29**

Name _____   Date _____

Diagnostic Test 1A

1. Write the fraction in decimal form.

$\frac{2}{4}$ _____

2. Write the mixed number as an improper fraction.

$1\frac{7}{10}$ _____

3. Do the fraction and the decimal match? Write yes or no.

$\frac{1}{10}$ and 0.01 _____

Score ⟋3

Circle one.
Minimal (0/3–1/3)
Basic (2/3)
Secure (3/3)

- -

Chapter 6

Name _____   Date _____

Diagnostic Test 2A

1. Write the fraction in decimal form.

$-\frac{2}{5}$ _____

2. Write the decimal in fraction form.

$-0.66666\ldots$ _____

3. Write whether the following number sentence is true or false.

$-\frac{9}{3} = \frac{9}{3}$

_____

Score ⟋3

Circle one.
Minimal (0/3–1/3)
Basic (2/3)
Secure (3/3)

Name _____ Date _____

Diagnostic Test 3A

1. Label the missing numbers on the number line.

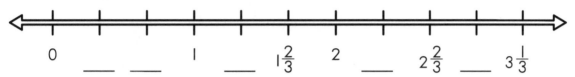

0 ___ ___ 1 ___ $1\frac{2}{3}$ 2 ___ $2\frac{2}{3}$ ___ $3\frac{1}{3}$

2. Estimate where each fraction is on the number line. Label each with a point and the fraction given.

$\frac{1}{10}$ and $\frac{1}{5}$

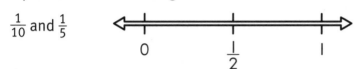

0     $\frac{1}{2}$ 1

3. Name a decimal between the given decimals. Record it on the number line.

Score ◿ 3

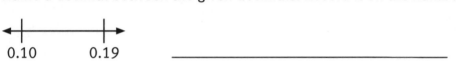

0.10   0.19   _____

Circle one.
Minimal (0/3–1/3)
Basic (2/3)
Secure (3/3)

Name _____ Date _____

Diagnostic Test 4A

1 Write the fraction that is located at the same point on a number line as the following decimal.

0.125   _____

2. Estimate where each decimal is on the number line. Label each with a point and the decimal given.

0.15 and 0.7

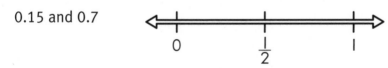

0     $\frac{1}{2}$ 1

3. Name a decimal between the given decimals. Record it on the number line.

Score ◿ 3

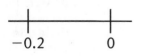

−0.2   0   _____

Circle one.
Minimal (0/3–1/3)
Basic (2/3)
Secure (3/3)

Name _____ Date _____

Diagnostic Test 1A

1. Write the factors in the expression.

   $12 \times 50 = 600$ _____ _____

2. Write the first five multiples of the number.

   4 _____

3. Write whether the following number is prime or composite.

   49 is a _____ number.

Score ◺ 3

Circle one.
Minimal (0/3–1/3)
Basic (2/3)
Secure (3/3)

· ·

Chapter
7

Name _____ Date _____

Diagnostic Test 2A

1. Write the following number as the product of its prime factors. Use counters, factor trees, mental math, or paper and pencil to determine the prime factors for the number.

   16 _____

2. Is a certain number a prime number evenly divisible by 2 and 5 if it is?

   _____

3. What are the prime factors of 40?

   _____

Score ◺ 3

Circle one.
Minimal (0/3–1/3)
Basic (2/3)
Secure (3/3)

Chapter 7

Diagnostic Test 3A

Name _____ Date _____

1. Write the exponent.

$5 \times 5 \times 5 \times 5 = 5$ _____

2. Write the base with its exponent.

$7 \times 7 \times 7 \times 7 \times 7 =$ _____

3. Write the following number as the product of its prime factors in expanded form and using exponents.

$98 =$ _____ $=$ _____

Score ⟋ 3

Circle one.
Minimal (0/3–1/3)
Basic (2/3)
Secure (3/3)

··

Chapter 7

Diagnostic Test 4A

Name _____ Date _____

1. How would you write 10 to the third power as repeated multiplication?

_____

2. Write the number in expanded form, and then in expanded form using powers of ten.

$910.5 =$ _____

$=$ _____

3. Write the number in standard form.

$5 \times 10^0 + 3 \times 10^{-1} + 2 \times 10^{-2} + 1 \times 10^{-3} =$ _____

Score ⟋ 3

Circle one.
Minimal (0/3–1/3)
Basic (2/3)
Secure (3/3)

Chapter 8

Name _____ Date _____

Diagnostic Test 1A

1. Do these models represent equivalent fractions? Write yes or no.

_____

$\frac{2}{5}$ $\frac{3}{8}$

2. Write the numerator or the denominator to make equivalent fractions.

$\frac{1}{4} = \frac{\square}{8} = \frac{3}{\square} = \frac{\square}{\square}$

3. Compare the shaded parts of each of these fractions, and write a true statement using <, >, or =.

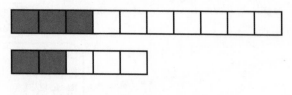

_____ _____ _____

Score /3

Circle one.
Minimal (0/3–1/3)
Basic (2/3)
Secure (3/3)

- -

Chapter 8

Name _____ Date _____

Diagnostic Test 2A

1. Use mental math or play money to find the answer to the question. Write your answer.

How many pennies are in $\frac{1}{2}$ of 100? _____

2. Write the pennies as a part of 100 in fraction form and decimal form.

51 pennies   _____ _____

3. Write the fraction as a decimal.

$\frac{912}{1000}$ _____

Score /3

Circle one.
Minimal (0/3–1/3)
Basic (2/3)
Secure (3/3)

Name _____ Date _____

Diagnostic Test 3A

1. Shade the picture below to represent each percentage given.

Shade 60% of the circle.

2. Write the answer to each question. You can draw models to help find percentages.

25% of 16

25% is the same as the fraction: _____.

The fraction, _____, of 16 is _____.

25% of 16 is _____.

3. Complete the chart below so the percent, fraction, and decimal equal each other.

Percent	Fraction	Decimal
35%		

Score / 3

Circle one.
Minimal (0/3–1/3)
Basic (2/3)
Secure (3/3)

- -

Name _____ Date _____

Diagnostic Test 4A

1. Label each missing number on the number line.

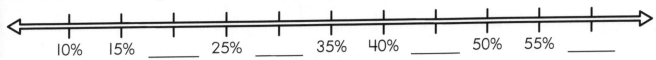

10%   15%   ____   25%   ____   35%   40%   ____   50%   55%   ____

2. Estimate where each number is on the number line. Label each with a point and the given value.

0.1 and 0.9

0          $\frac{1}{2}$ 1

3. Name a percentage between the given values. Record it on the number line.

$\frac{1}{10}$ $\frac{1}{2}$ _____

Score / 3

Circle one.
Minimal (0/3–1/3)
Basic (2/3)
Secure (3/3)

Name _____ Date _____

Diagnostic Test 1A

1. Write what direction you move on a number line when you subtract a

   positive number. _____

2. Use the number line to add the numbers. Circle the starting number.
   Use arrows to show the amount added. Put a square around the answer.

   $-10 + 4 =$ _____

   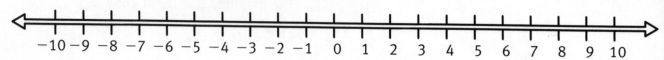

3. Use the number line to help you solve the problem.

   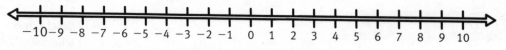

   $5 - -4 =$ _____

   Score ◰ /3

   Circle one.
   Minimal (0/3–1/3)
   Basic (2/3)
   Secure (3/3)

- -

Name _____ Date _____

Diagnostic Test 2A

Use each grid to find the sums. Write the solution, and fill in the grid.

1. $\frac{1}{5} + \frac{2}{5} =$ _____

2. $\frac{4}{8} - \frac{1}{8} =$ _____

3. $-\frac{1}{2} + \frac{5}{6} =$ _____

   Score ◰ /3

   Circle one.
   Minimal (0/3–1/3)
   Basic (2/3)
   Secure (3/3)

Diagnostic Test 3A

Use each grid to add of subtract the decimals.

1. $0.70 - 0.17 =$ _____

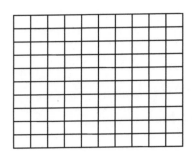

2. $0.28 + 0.41 =$ _____

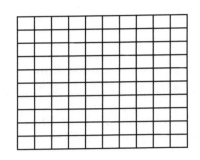

3. $0.24 + -0.17 =$ _____

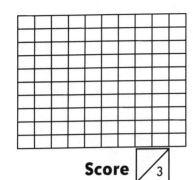

Score ⧄ 3

Circle one.
Minimal (0/3–1/3)
Basic (2/3)
Secure (3/3)

· ·

Diagnostic Test 4A

Decide whether each equation is true or false. Explain. Write your answers.

1. $1^{13} = 113$

The statement is _____ because _____

2. $45^0 = 0$

The statement is _____ because _____

3. Write the solution to the equation.

$17^2 =$ _____

Score ⧄ 3

Circle one.
Minimal (0/3–1/3)
Basic (2/3)
Secure (3/3)

Name _____ Date _____

Diagnostic Test 1A

1. Is zero an integer? _____

2. $-17 \times -20 =$ _____

3. Decide whether to multiply or divide in the following situation. Explain why you chose the operation you chose. Write your answers.

   A designer is trying to buy enough wallpaper for an 8-foot by 14-foot wall. How many feet of wallpaper does he need to buy?

   8 ___ 14

   I chose this operation because

   _____

Score ⧄ /3

Circle one.
Minimal (0/3–1/3)
Basic (2/3)
Secure (3/3)

• •

Name _____ Date _____

Diagnostic Test 2A

1. Model and write the answer for the multiplication problem.

   $\frac{1}{3} \times \frac{1}{5} =$ _____

   A model for the multiplication problem is:

2. Reduce the fraction by dividing the numerator and the denominator by a common factor. Write your answer.

   $\frac{36}{24} =$ _____

   The common factor I divided the numerator

   and denominator by was _____.

3. Divide the fractions. Write your answer as a reduced fraction.

   $\frac{3}{8} \div \frac{6}{2} =$ _____

Score ⧄ /3

Circle one.
Minimal (0/3–1/3)
Basic (2/3)
Secure (3/3)

Name _____ Date _____

Diagnostic Test 3A

1. If you were multiplying 0.253 by 0.1, how many numbers would be to the right of the decimal point? _____

2. Use the grid to multiply the decimals.

   $0.5 \times 0.8 =$ _____

   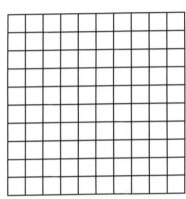

3. Solve the problem. Show your work, and write your answer.

   A bird is traveling 63.1 miles per hour. How far will the bird travel over a period of

   5.5 hours? _____ miles

   Score 

   Circle one.
   Minimal (0/3–1/3)
   Basic (2/3)
   Secure (3/3)

Name _____ Date _____

Diagnostic Test 4A

1. Multiply the decimal by the power of ten. Write the product.

   $10 \times 54.6 =$ _____

2. Divide the expression. Write the result.

   $0.72 \div 0.024 =$ _____

3. Multiply the expression. Write the result.

   $0.016 \times 0.32 =$ _____

   Score

   Circle one.
   Minimal (0/3–1/3)
   Basic (2/3)
   Secure (3/3)

Name _____ Date _____

Diagnostic Test 1A

Complete the tables below so the percent, fraction, and decimal in each row equal each other.

1.

Percent	Fraction
45%	

2.

Percent	Decimal
	0.04

3.

Fraction	Decimal
$\frac{1}{8}$	

Score ⬜/3

Circle one.
Minimal (0/3–1/3)
Basic (2/3)
Secure (3/3)

Chapter 11

Name _____ Date _____

Diagnostic Test 2A

1. Write two different fractions that are equivalent to $\frac{7}{8}$.

_____

Convert each decimal into a fraction. Reduce the fraction to lowest terms. Write your answers.

2. $0.107 =$ _____

reduced form is _____

3. $0.32 =$ _____

reduced form is _____

Score /3

Circle one.
Minimal (0/3–1/3)
Basic (2/3)
Secure (3/3)

Name _____ **Date** _____

Diagnostic Test 3A

There are 30 people in Jo's class. Sixty percent of the students are in eleventh grade. How many people are in eleventh grade?

1. What fraction would you set up to represent the percent?

   _____

2. What fraction would you set up to represent the people in eleventh grade (the part) divided by total number of people in the class (the whole)?

   _____

3. What is the proportional equation that you set up? How many people are in eleventh grade?

   _____ _____

Score ⬜ /3

Circle one.
Minimal (0/3–1/3)
Basic (2/3)
Secure (3/3)

Name _____ **Date** _____

Diagnostic Test 4A

School supplies are on sale for 40% off. If a backpack is normally $30, then what is the price after the discount?

1. What proportion should you set up to find the amount that is discounted? What is the amount that is discounted?

   _____ _____

2. What is the subtraction problem you should set up to find the final price?

   _____

3. What is the final price? _____

Score ⬜ /3

Circle one.
Minimal (0/3–1/3)
Basic (2/3)
Secure (3/3)

Name _____ Date _____

Diagnostic Test 1A

1. Write the following multiplication problem as an exponential expression.

   $3 \times 3 \times 3 \times 3 \times 3 =$ _____

2. In expanded form, 9^3 is _____.

3. Simplified into standard form, 9^3 is _____.

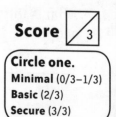

Score ⟋ 3

Circle one.
Minimal (0/3–1/3)
Basic (2/3)
Secure (3/3)

Name _____ Date _____

Diagnostic Test 2A

1. Write the following division problem as an exponential expression.

   $1 \div 15 \div 15 \div 15 \div 15 =$ _____

2. In expanded form, 3^{-3} is _____.

3. Simplified into standard form, 3^{-3} is _____.

Score ⟋ 3

Circle one.
Minimal (0/3–1/3)
Basic (2/3)
Secure (3/3)

Name _____ Date _____

Diagnostic Test 3A

1. Change the expression with a negative exponent into a fraction with a positive exponent in the denominator.

$5^{-4} =$ _____

2. Write the expression modeled by the Fraction Tiles using a negative exponent. Rewrite the expression as a fraction with a positive exponent.

_____ = _____

3. Write 5^{-6} using repeated division.

_____

Score ⬦ 3

Circle one.
Minimal (0/3–1/3)
Basic (2/3)
Secure (3/3)

Name _____ Date _____

Diagnostic Test 4A

1. Rewrite the exponential expression using repeated multiplication. Cross out an equal number of factors from the top and bottom. Write the resulting exponent.

$8^5 \div 8^2 =$ _____ = _____

2. Rewrite the exponential expression using repeated multiplication. Write the resulting exponent.

$3^5 \times 3^3 =$ _____ = _____

3. Divide. Write the result as an exponential expression.

$4^3 \div 4^{-2} =$ _____

Score ⬦ 3

Circle one.
Minimal (0/3–1/3)
Basic (2/3)
Secure (3/3)

Name _____ Date _____

Diagnostic Test 1A

1. Evaluate the expression by completing the operation inside the parentheses first. Circle the correct answer.

$3 \times (5 + 12) =$ 27   51

2. Draw parentheses around the operation that should be performed first to get the correct answer.

$11 + 2 \times 4 = 19$

3. Write the solution to the expression.

$7 \times (2 + 10) =$ _____

Score /3

Circle one.
Minimal (0/3–1/3)
Basic (2/3)
Secure (3/3)

Name _____ Date _____

Diagnostic Test 2A

1. Which operation should you perform first?

$(2 + 3) \times 5$ _____

2. Use the digits to reach the target number. Write each digit one time in the correct space.

Target number: 72

Digits to combine: 0, 4, 5, 8

Equation:

(_____ + _____) × (_____ + _____)

3. Use the operations to reach the target number. Write each operation one time in the correct space.

Target number: 19

Operations to combine: ×, +, +

Equation: (1 _____ 2) _____ (9 _____ 8)

Score /3

Circle one.
Minimal (0/3–1/3)
Basic (2/3)
Secure (3/3)

Name _____ Date _____

Diagnostic Test 3A

1. Write the first operation you would perform in the equation.

   $11 + 3 \div 4$

   The first operation is _____.

2. Write the first operation you would perform in the equation.

   $5^{-4} + 16$

   The first operation is _____.

Score ⬚/3

Circle one.
Minimal (0/3–1/3)
Basic (2/3)
Secure (3/3)

3. Evaluate the following expression. Write the solution.

   $2 + (5 \times 1)^2 \times 2 =$ _____

Name _____ Date _____

Diagnostic Test 4A

1. Use Counters to model the expression that is described.

   If you add 5 to a number, you get $\frac{1}{4}$ of 40. What is the number?

2. Write the expression that is described.

   If you subtract 7 from a number, you get 3 times 12. What is the number?

   _____

3. Write the solution to the expression that is described.

   If you add 8 to a number, you get 5 times 9. What is the number?

   _____

Score ⬚/3

Circle one.
Minimal (0/3–1/3)
Basic (2/3)
Secure (3/3)

Name _____ Date _____

Diagnostic Test 1A

1. Write the evaluation of the expression with the variable as defined.

Expression: $5d$ Evaluation: _____

Variable: $d = 6$

2. Write the expression that would give you the outputs desired.

Input	Output
0	0
1	−2
2	−4
3	−6

Expression: _____

3. Write the expression that would give you the outputs desired.

Input	Output
0	7
1	8
2	9
3	10

Expression: _____

Score ◸/3

Circle one.
Minimal (0/3–1/3)
Basic (2/3)
Secure (3/3)

Name _____ Date _____

Diagnostic Test 2A

1. Solve for the variable. It might be useful to use Counters to model the situation.

$20 + a = 19$ _____

3. Complete the table by writing the missing input or output for the expression $8x$.

Input	Output
$x = -1$	
	48

2. Solve for the variable. It might be useful to use Counters to model the situation.

$8b = 40$ _____

Score ◸/3

Circle one.
Minimal (0/3–1/3)
Basic (2/3)
Secure (3/3)

Name _____ Date _____

Diagnostic Test 3A

1. What operation is the word *product* associated with?

_____

2. Draw a number line to represent the quantity described by the expression.

10 divided by a number

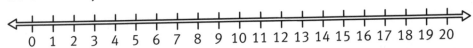

3. Write the expression that is described.

The sum of 11 and a number   _____

Score [/3]

Circle one.
Minimal (0/3–1/3)
Basic (2/3)
Secure (3/3)

Name _____ Date _____

Diagnostic Test 4A

1. Use the number line to represent the quantity described in the expression. Write the expression.

A number is less than 5.

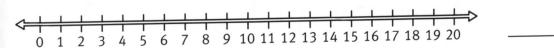   _____

2. Draw a number line to represent the quantity described in the expression. Write the expression.

A number is less than or equal to 5.

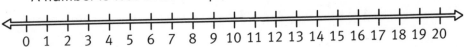   _____

Score [/3]

3. Write the equation that is described.

A number *y* is equal to *x* to the third power.   _____

Circle one.
Minimal (0/3–1/3)
Basic (2/3)
Secure (3/3)

Name _____ Date _____

Diagnostic Test 1A

1. Write the *y*-value in the table for the equation $y = 2 + 4x$.

x	y
−1	
3	

2. Create a table to represent some solutions for the following two-variable equation.

$y = x + 4$

x	y

3. Create a table to represent some solutions for the following two-variable equation.

$y = -x$

x	y

Score ◻/3

Circle one.
Minimal (0/3–1/3)
Basic (2/3)
Secure (3/3)

Name _____ Date _____

Diagnostic Test 2A

1. If the first scale is balanced, will the second scale be balanced? Write *yes* or *no*.

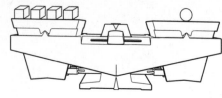 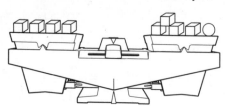

_____

2. Write the number that makes the statement true.

$z = g; z + 3 = g +$ _____

3. Write the number that makes the statement true.

$j = p; j - 1 = p -$ _____

Score ◻/3

Circle one.
Minimal (0/3–1/3)
Basic (2/3)
Secure (3/3)

Name _____ Date _____

Diagnostic Test 3A

1. If the first scale is balanced, will the second scale be balanced?
Write *yes* or *no*.

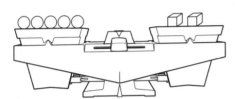 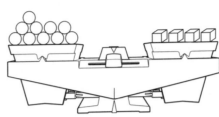

_____

2. Write the number that makes the statement true.

$a = b$; $5a = $ _____ b

Score ⬚/3

Circle one.
Minimal (0/3–1/3)
Basic (2/3)
Secure (3/3)

3. Write the number that makes the statement true.

$c = d$; $10c = ($ _____ $+ 3)d$

. . . .

Name _____ Date _____

Diagnostic Test 4A

1. Write the new equation that results after performing the indicated
operations to both sides of the equation.

$5x = 20$; divide by 5.   _____

2. Write an operation to help you isolate the variable, and then
perform that operation on both sides. Write the resulting equation.

$-7x = 56$

The operation I performed is _____.

The resulting equation is _____.

Score ⬚/3

3. Isolate the variable. Write the solution.

$x + 5 = 2$ _____

Circle one.
Minimal (0/3–1/3)
Basic (2/3)
Secure (3/3)

Chapter 16

Name _____ Date _____

Diagnostic Test 1A

1. Write which property you can use to simplify the expression.

   $6 \times \frac{1}{6}$ _____

2. Apply the property to simplify the expression. Write the resulting expression.

   $3 \times (5 + 2)$; distributive property

   _____

3. Apply the property to simplify the expression. Write the resulting expression.

   $15 \times (7 \times 8)$; associative property

   _____

   Score /3

- -

Chapter 16

Name _____ Date _____

Diagnostic Test 2A

1. Write the two steps you need to perform in the following two-step equation.

   $2x + 4 = 10$ _____

2. Write the solution to the two-step equation.

   $-x + 2 = 5$

   _____

3. Verify your solution by substituting for x. Show your work.

   $-\underline{\quad} + 2 = 5$

   _____

   Score /3

Name _____ Date _____

Diagnostic Test 3A

1. Write the two steps you need to perform in the following two-step equation.

   $x \div 4 + 1 < 10$

   _____

2. Write the solution to the two-step equation.

   $6x + 4 \geq 10$

   _____

3. Verify your solution by substituting for x. Show your work.

   $6\underline{} + 4 \geq 10$

   _____

Score ⬚ / 3

Circle one.
Minimal (0/3–1/3)
Basic (2/3)
Secure (3/3)

Name _____ Date _____

Diagnostic Test 4A

Someone eats a total of 15 ounces of vegetables for dinner in 5 days, and 20 ounces of vegetables for lunch in 5 days. At these rates, how many ounces of vegetables does he eat altogether in 9 days?

1. What rates do you need to find?

   _____

   _____

   _____

2. What are these rates?

   _____

   _____

   _____

3. How do you use these rates to find out how many ounces of vegetables he ate for dinner over 9 days, and how many ounces of vegetables he ate for lunch over 9 days?

   _____

Score ⬚ / 3

Circle one.
Minimal (0/3–1/3)
Basic (2/3)
Secure (3/3)

Chapter 17

Name ______________________ Date __________

Diagnostic Test 1A

1. Which coordinate comes first when writing an ordered pair?

____

2. Give the coordinates of the point indicated by the letter. Be sure to use correct notation.

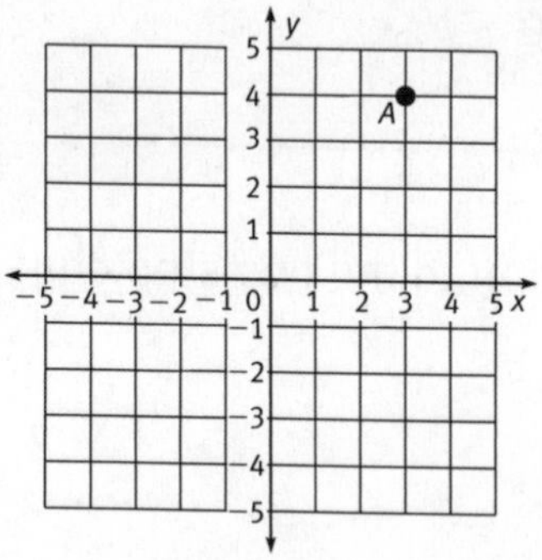

A ______

3. How would you move from (5, 2) to (7, 3)? Use a coordinate grid to help you answer the question. Be sure to describe left-right movement first and then up-down movement.

____

Score /3

Circle one.
Minimal (0/3–1/3)
Basic (2/3)
Secure (3/3)

Chapter 17

Name ______________________ Date __________

Diagnostic Test 2A

1. What can you say about the first coordinate in any point in Quadrant IV?

____

2. Write the quadrant in which the point is located. You can use a coordinate grid to help you identify the location. If the point is not in a quadrant, list the axis or origin on which it lies.

(0, 8) ______________________

3. Use a coordinate grid to graph (2, −5). Label the point using its coordinates. Be sure to use correct punctuation.

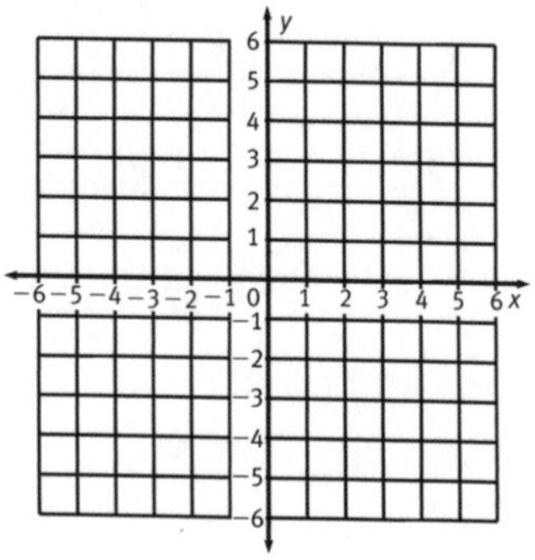

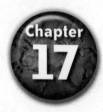

Score /3

Circle one.
Minimal (0/3–1/3)
Basic (2/3)
Secure (3/3)

Name _____ Date _____

Diagnostic Test 3A

1. Use a ruler to draw a line that goes through each pair of points on these coordinate grids.

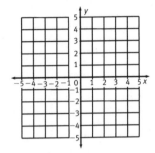

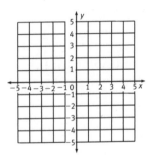

2. Plot (5, 5) on the coordinate grid. Use a ruler to draw a line that goes through the origin and the point you have plotted.

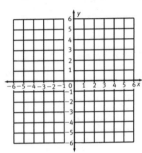

3. Plot (0, 4) and (2, 3) on the coordinate grid. Use a ruler to draw a line that goes through the two points you have plotted.

Score ◻ / 3

Circle one.
Minimal (0/3–1/3)
Basic (2/3)
Secure (3/3)

Name _____ Date _____

Diagnostic Test 4A

1. Graph the following points on the coordinate plane: *A* (5, 7); *B* (1, 7); *C* (1, 1); *D* (5, 1); *E* (3, 4); and *F* (1, 4).

2. Connect points *A*, *B*, *C*, and *D* in alphabetical order. Stop, and pick up your pencil. Now connect Point *E* to Point *F*.

3. Identify the shape you have made.

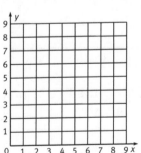

The shape is a(n)

_____

Score ◻ / 3

Circle one.
Minimal (0/3–1/3)
Basic (2/3)
Secure (3/3)

Chapter 18

Name _____ Date _____

Diagnostic Test 1A

1. Which coordinate is associated with horizontal distance?

   _____

2. Determine the length of the line segment that connects $(-5, 2)$ and $(-3, 2)$. Show your work.

   _____

3. Determine the horizontal distance between the endpoints of the line segment that connects $(0, 5)$ and $(2, 9)$. Show your work.

   _____

Score ⬜/3

Circle one.
Minimal (0/3–1/3)
Basic (2/3)
Secure (3/3)

· ·

Chapter 18

Name _____ Date _____

Diagnostic Test 2A

1. Which coordinate is associated with vertical distance?

   _____

2. Determine the length of the line segment that connects $(5, 4)$ and $(5, 10)$. Show your work.

   _____

3. Determine the vertical distance between the endpoints of the line segment that connects $(0, 5)$ and $(2, 9)$. Show your work.

   _____

Score ⬜/3

Circle one.
Minimal (0/3–1/3)
Basic (2/3)
Secure (3/3)

Diagnostic Test 3A

1. Find the length of side *KL*. Call it *a* in the Pythagorean theorem. Show your work.

_____

2. Find the length of side *KM*. Call it *b* in the Pythagorean theorem. Show your work.

_____

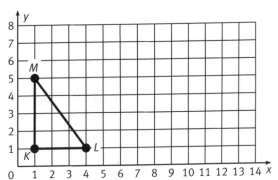

3. How long is the hypotenuse, side *ML*, in this right triangle?

_____

Score /3

Circle one.
Minimal (0/3–1/3)
Basic (2/3)
Secure (3/3)

- -

Chapter 18

Name _____ Date _____

Diagnostic Test 4A

1. What is a Pythagorean triple?

_____

2. Give an example of a Pythagorean triple.

_____

3. Consider the following group of three numbers, and determine whether they are the sides in a right triangle. Explain your answer. You will use the Pythagorean theorem. The hypotenuse will be identified by *c*.

9; 12; *c* = 13

_____

Score /3

Circle one.
Minimal (0/3–1/3)
Basic (2/3)
Secure (3/3)

Name _____ Date _____

Diagnostic Test 1A

1. What is the change in *y*-value for the pair of points (2, 8) and (−3, 7)?

_____

2. What is the change in *x*-value for the pair of points (2, 8) and (−3, 7)?

_____

3. What is the slope of these two points?

_____

Score /3

Circle one.
Minimal (0/3–1/3)
Basic (2/3)
Secure (3/3)

- -

Name _____ Date _____

Diagnostic Test 2A

1. Use the formula for the slope given two points, and find the slope of the line through the pair of points (−4, 3) and (−6, 7).

The slope of this line is _____

2. The slope of this line is

_____.

3. The slope of this line is _____.

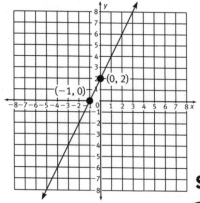

Score /3

Circle one.
Minimal (0/3–1/3)
Basic (2/3)
Secure (3/3)

Name _____ Date _____

Diagnostic Test 3A

1. One minute equals 60 seconds. Write an equation to express this relationship using *x* and *y*.

   _____

2. From this equation, three ordered pairs are _____, _____, and _____.

3. Graph this direct-variation relationship.

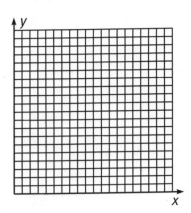

Score ☐/3

Circle one.
Minimal (0/3–1/3)
Basic (2/3)
Secure (3/3)

Name _____ Date _____

Diagnostic Test 4A

1. The width of an object is three times its length. Write an equation to express this relationship using *x* and *y*.

   _____

2. From this equation, three ordered pairs are _____, _____, and _____.

3. Graph this direct-variation relationship.

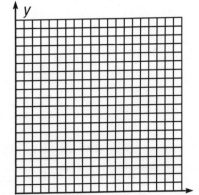

Score /3

Circle one.
Minimal (0/3–1/3)
Basic (2/3)
Secure (3/3)

Name _____ Date _____

Diagnostic Test 1A

1. What is the slope of the line $y = 2x + 5$?

   _____

2. What is the y-interecept of the line $y = 2x + 5$?

   _____

3. Graph the line for the equation.

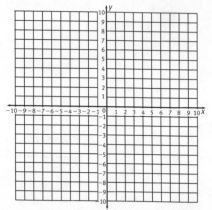

Score ⟋ 3

Circle one.
Minimal (0/3–1/3)
Basic (2/3)
Secure (3/3)

Name _____ Date _____

Diagnostic Test 2A

1. Joanna earns $8 per hour. If she works for 10 hours, how much money will she earn?

   _____

2. If she wants to earn $240, how many hours does she need to work?

   _____

3. How many fewer hours could she work to make $240 if she made $10 per hour?

   _____

Score 3

Circle one.
Minimal (0/3–1/3)
Basic (2/3)
Secure (3/3)

Diagnostic Test 3A

It is time to drain the swimming pool in the apartment complex where Anton lives. If the owners open 4 drains, it will take 20 hours to drain the pool.

1. How many faucet-hours are needed for this project?

   _____

2. How many hours would it take if the owners opened 10 drains?

   _____

3. Cold weather is coming, and they want to have the pool drained by tomorrow morning. How many drains do they need to open to get the pool drained in 10 hours?

   _____

Score ⬜/3

Circle one.
Minimal (0/3–1/3)
Basic (2/3)
Secure (3/3)

Diagnostic Test 4A

The speed limit on many highways is 55 miles per hour. At this rate, how many feet per minute does a car travel? Follow these steps to help you find the answer.

1. Convert miles to feet. *Miles* is on top in the first ratio, so you want *miles* on the bottom of this ratio to cancel out *miles*.

   _____

2. Convert hours to minutes. *Hour* is on the bottom in the first ratio, so you want *hour* on the top in this ratio to cancel out *hour*.

   _____

3. Use dimensional analysis to cancel out the units. The original question was *feet per second*, so you want to have *feet* left in the top and *seconds* left in the bottom.

   _____

   _____

   The result is _____ feet per minute.

Score ⬜/3

Circle one.
Minimal (0/3–1/3)
Basic (2/3)
Secure (3/3)

Name _____ Date _____

Diagnostic Test 1A

1. What is the opposite of 5?

   _____

2. What is the reciprocal of −7?

   _____

3. Why do we need to know how to find reciprocals?

   _____

   _____

Score ⧄ 3

Circle one.
Minimal (0/3–1/3)
Basic (2/3)
Secure (3/3)

Chapter 21

Name _____ Date _____

Diagnostic Test 2A

1. The following statement is either true or false because the distributive property has been correctly or incorrectly applied. Decide whether the statement is true or false by substituting the given number for the variable.

   $-3(x - 4) = -3x - 4$; let $x = 1$

   _____

   _____

   _____

2. Simplify the following expression using the distributive property.

   $3(x + 5)$ _____

3. Why do we use the distributive property?

   _____

   _____

   _____

Score ⧄ 3

Circle one.
Minimal (0/3–1/3)
Basic (2/3)
Secure (3/3)

Name _____ Date _____

Diagnostic Test 3A

1. What is the simplification of the following root?

   $\sqrt[3]{125}$

   _____

2. What is the simplification of the following exponent?

   13^2

   _____

3. What is the inverse operation of the following expression?

   $\sqrt{49} = 7$

   _____

Score [/3]

Circle one.
Minimal (0/3–1/3)
Basic (2/3)
Secure (3/3)

Chapter 21

Name _____ Date _____

Diagnostic Test 4A

1. What is the simplification of the following expression?

   $100^{\frac{1}{2}}$

   _____

2. How would you write the following expression as a root?

   $23^{\frac{1}{4}}$

   _____

3. What is the inverse exponential expression of the following expression?

   $7^2 = 49$

   _____

Score [/3]

Circle one.
Minimal (0/3–1/3)
Basic (2/3)
Secure (3/3)

Diagnostic Test 1A

1. Use the Pythagorean theorem to decide whether the following sides form a right triangle.

   4, 7, 9

   _____

Use the Pythagorean theorem to find the length of the missing sides.

2. $a = 5$; $b = 3$; $c =$ _____

3. $a = 10$; $b =$ _____; $c = 12$

Score ⬜ /3

Circle one.
Minimal (0/3–1/3)
Basic (2/3)
Secure (3/3)

Chapter 22

Name _____ Date _____

Diagnostic Test 2A

Simplify the following expressions.

1. $8^{-\frac{1}{3}}$

   _____

2. 20^0

   _____

3. $(3x)^2$

   _____

Score ⬜ /3

Circle one.
Minimal (0/3–1/3)
Basic (2/3)
Secure (3/3)

Chapter 22

Name _____ Date _____

Diagnostic Test 3A

Simplify the following expressions.

1. $x^3 \times x^4$

   _____

2. $x^4 \div x^{10}$

   _____

3. $(y^3)^3$

   _____

Score ⬜ /3

Circle one.
Minimal (0/3–1/3)
Basic (2/3)
Secure (3/3)

· ·

Chapter 22

Name _____ Date _____

Diagnostic Test 4A

Simplify the expression using the 5 Basic Rules, PEMDAS, the distributive property, and the rules for signed numbers. Assume variables do not equal 0.

1. $-2x^2 + 10x^2 =$ _____

2. $(-4x)^2 =$ _____

Score ⬜ /3

Circle one.
Minimal (0/3–1/3)
Basic (2/3)
Secure (3/3)

3. $-x^4(-3x - 4) =$ _____

Name _____ **Date** _____

Diagnostic Test 1A

Expand the following expressions.

1. $9x^4 \div 6x^2$

   _____

2. $9x^4 \div 6x^2$

   _____

3. $(\frac{1}{4})(16x^2 - 4x) =$ _____

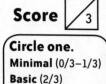

Score ⬛/3

Circle one.
Minimal (0/3–1/3)
Basic (2/3)
Secure (3/3)

Name _____ **Date** _____

Diagnostic Test 2A

Write the property or rule used to get from the first equation to the second equation.

1. $3y - 7 = 20$

   $3y - 7 + 7 = 20 + 7$

   _____

2. $20y = 100$

   $\frac{20y}{20} = \frac{100}{20}$

   _____

3. $2(y + 3) = 8$

   $2y + 6 = 8$

   _____

Score ⬛/3

Circle one.
Minimal (0/3–1/3)
Basic (2/3)
Secure (3/3)

Name _____ Date _____

Diagnostic Test 3A

1. Write an example of the distributive property.

   _____

2. Write an example of the associative property.

   _____

3. Write an example of the commutative property.

   _____

Score ⬚/3

Circle one.
Minimal (0/3–1/3)
Basic (2/3)
Secure (3/3)

· ·

Name _____ Date _____

Diagnostic Test 4A

Solve the equations, and prove your answers.

1. $9 = c + 5 - 2c + 4$

   $c = \underline{\hspace{1cm}}$

2. $3n - 5(n - 2) + 4n = 2(9 + 3)$

   $n = \underline{\hspace{1cm}}$

3. $\frac{1}{2}(4y - 2) = 2^3 + 3$

   $y = \underline{\hspace{1cm}}$

Score ⬚/3

Circle one.
Minimal (0/3–1/3)
Basic (2/3)
Secure (3/3)

Name _____ Date _____

Diagnostic Test 1A

Translate the verbal descriptions into equations, and solve.

1. Three more than half a number is 4.

   _____

2. The difference between a positive number and 20 is 31.

   _____

3. The sum of a number and 14 is 46.

   _____

Score [/3]

Circle one.
Minimal (0/3–1/3)
Basic (2/3)
Secure (3/3)

Name _____ Date _____

Diagnostic Test 2A

Draw number lines for the following inequalities.

1. $x \geq 2$

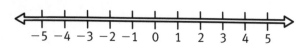

2. $x < 2$

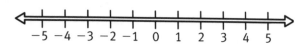

3. $x = 2$

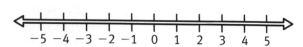

Score [/3]

Circle one.
Minimal (0/3–1/3)
Basic (2/3)
Secure (3/3)

Name _____ Date _____

Diagnostic Test 3A

The local zoo has two pricing structures. The Family Plan costs $35 for a family of 4, and $5 for each additional family member. The Individual Ticket Plan costs $12 per person. How large would a family need to be so that the Family Plan price is a better buy?

1. What are the variables?

   _____

2. What are the equations in terms of these variables?

   _____

Score ⬚/3

Circle one.
Minimal (0/3–1/3)
Basic (2/3)
Secure (3/3)

3. How would you set up the inequality?

   _____

Name _____ Date _____

Diagnostic Test 4A

1. Write an example of an expression.

   _____

2. Write an example of an equation.

   _____

3. Write an example of an inequality.

   _____

Score /3

Circle one.
Minimal (0/3–1/3)
Basic (2/3)
Secure (3/3)

Name _____ Date _____

Test

Fill in the blanks to show the value of each digit.

1. 2,548   _____ × 1,000 = _____

            _____ × 100 = _____

            _____ × 10 = _____

            _____ × 1 = _____

3. 3,084   _____ × 1,000 = _____

            _____ × 100 = _____

            _____ × 10 = _____

            _____ × 1 = _____

2. 4,894   _____ × 1,000 = _____

            _____ × 100 = _____

            _____ × 10 = _____

            _____ × 1 = _____

4. 9,275   _____ × 1,000 = _____

            _____ × 100 = _____

            _____ × 10 = _____

            _____ × 1 = _____

Place <, >, or = in each blank.

5. 2,548 _____ 2,568

7. 208 _____ 208

6. 480 _____ 489

8. 9,275 _____ 3,275

Name _____ Date _____

Test

Find each sum. Write your answer.

9.   267
    + 824
    ‾‾‾‾‾‾

11.   241
    + 372
    ‾‾‾‾‾‾

10.   845
    + 277
    ‾‾‾‾‾‾

12.   292
    + 557
    ‾‾‾‾‾‾

Find each sum. Write your answer.

13. 527 + 807

_____;

17. 961 + 784

_____;

14. 287 + 913

_____;

18. 402 + 123

_____;

15. 229 + 564

_____;

19. 284 + 407

_____;

16. 480 + 57

_____;

20. 745 + 42

_____;

Name _____ Date _____

Test

Use the digits given to make the largest number possible. Use each digit one time. Write your answer.

1. 8, 1, 2, 8 _____

3. 0, 0, 9, 8 _____

2. 3, 2, 8, 5 _____

4. 5, 0, 4, 4 _____

Choose the two numbers from each group that have the greatest difference. Write and solve that number sentence.

5. 480   387   342

_____

9. 801   228   671

_____

6. 256   133   405

_____

10. 970   977   205

_____

7. 205   316   382

_____

11. 378   535   831

_____

8. 143   288   390

_____

12. 941   559   808

_____

Name _____ Date _____

Test

Find each difference. Write your answer. Use addition to check your answer.

13.
$$\begin{array}{r} 152 \\ -\ 33 \\ \hline \end{array}$$

Check:
$$\begin{array}{r} 33 \\ +\ \underline{} \\ \hline \end{array}$$

14.
$$\begin{array}{r} 671 \\ -\ 78 \\ \hline \end{array}$$

Check:
$$\begin{array}{r} \\ +\ \underline{} \\ \hline \end{array}$$

Write the missing digits in each subtraction problem.

15.
$$\begin{array}{r} 2\ 1\ \_ \\ -\ 1\ \_\ 8 \\ \hline 2\ 3 \end{array}$$

18.
$$\begin{array}{r} \_\ 2\ 4 \\ -\ 5\ 8\ \_ \\ \hline 1\ \_\ 9 \end{array}$$

16.
$$\begin{array}{r} 3\ 6\ \_ \\ -\ 2\ \_\ 5 \\ \hline 1\ 0\ 5 \end{array}$$

19.
$$\begin{array}{r} \_\ 2\ 0 \\ -\ 1\ 0\ \_ \\ \hline 2\ \_\ 2 \end{array}$$

17.
$$\begin{array}{r} \_\ 0\ 4 \\ -\ 1\ 0\ \_ \\ \hline 2 \end{array}$$

20.
$$\begin{array}{r} 6\ 0\ \_ \\ -\ 1\ \_\ 7 \\ \hline \_\ 0\ 5 \end{array}$$

Name _____ Date _____

Test

Find each product. Write your product as repeated addition.

1. 2 × 4

_____

_____

3. 5 × 6

_____

_____

2. 1 × 9

_____

_____

4. 7 × 3

_____

_____

Break the second factor of each problem into a sum. Find each product. Write your answer.

5. 8 × 38

(8 × _____) + (8 × _____) = _____

9. 10 × 52

(10 × _____) + (10 × _____) = _____

6. 7 × 24

(7 × _____) + (7 × _____) = _____

10. 30 × 43

(30 × _____) + (30 × _____) = _____

7. 7 × 67

(7 × _____) + (7 × _____) = _____

11. 12 × 97

(12 × _____) + (12 × _____) = _____

8. 4 × 19

(4 × _____) + (4 × _____) = _____

12. 20 × 45

(20 × _____) + (20 × _____) = _____

Name _____ Date _____

Test

Write the property shown in each equation.

9. $12 \times (15 \times 2) = (12 \times 15) \times 2$

_____

10. $50 \times (10 + 3) = (50 \times 10) + (50 \times 3)$

_____

11. $2 \times (80 + 4) = (2 \times 80) + (2 \times 4)$

_____

12. $15 \times (40 + 3) = (15 \times 40) + (15 \times 3)$

_____

13. $5 \times 70 = 70 \times 5$

_____

14. $80 \times (23 \times 4) = (80 \times 23) \times 4$

_____

Use the traditional method to find each product. Write your answer.
Show your work.

15. $\begin{array}{r} 24 \\ \times\ 12 \\ \hline \end{array}$

16. $\begin{array}{r} 82 \\ \times\ 47 \\ \hline \end{array}$

17. $\begin{array}{r} 341 \\ \times\ 19 \\ \hline \end{array}$

18. $\begin{array}{r} 297 \\ \times\ 26 \\ \hline \end{array}$

Chapter 4

Name _____ Date _____

Test

Solve each problem below using Counters or drawing circles. Write a division number sentence to describe each situation.

1. Forty people are going on a trip. They will be organized evenly into 5 vans. How many people will be in each van?

_____ people in each van _____

2. Coach Vasquez arranged 32 members of the track team into groups of 4 for a practice. How many groups of athletes were there?

_____ groups of athletes _____

Solve each problem below. Write a division number sentence to model each situation.

3. Jan is helping restore cars. She has 72 fuses for the 10 cars. How many fuses will each car have? How many fuses will be left over?

_____

_____

_____

5. Vadim has 425 CDs in his collection. If he wants to distribute his CDs evenly among 10 shelves, then how many CDs should be on each shelf? How many CDs are left over?

_____

_____

_____

4. A tailor puts 6 buttons on each shirt. He purchases the buttons in packages. Each package has 40 buttons. How many shirts can the tailor make using 1 package of buttons? How many buttons are left over?

_____

_____

_____

6. Flora is obtaining dog treats for the local animal rescue. Each dog will require 2 biscuits and 4 bones. If the local store donated 50 biscuits and 80 bones, how many dogs can Flora give the treats to? Explain.

_____

_____

_____

_____

Use long division to find each quotient. Write your answer. Show your work.

7. $7\overline{)883}$

9. $12\overline{)508}$

8. $8\overline{)734}$

10. $14\overline{)812}$

Solve each problem. Decide how to handle each remainder. Write your answer.

11. Jameson has $25 to buy board games. Board games are on sale for $8 each. How many board games can Jameson buy?

_____ board games

What did you do with the remainder? Explain.

_____

_____

12. Valla has a set of 54 collectible cars. She wants to store them on shelves that hold 10 cars each. How many shelves will she need to store the cars?

_____ shelves

What did you do with the remainder? Explain.

_____

_____

Test

Write the fraction that represents the shaded area. Do not reduce the fractions.

1. 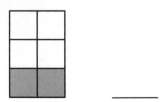 _____

3. 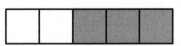 _____

2. 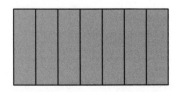 _____

4.  _____

Divide each shape into the number of equal parts indicated by the denominator. Then shade the unit fraction as indicated by the numerator. Write the fractional amount that is not shaded.

5. $\frac{1}{12}$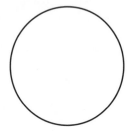

_____ is not shaded.

7. $\frac{1}{2}$

_____ is not shaded.

6. $\frac{1}{5}$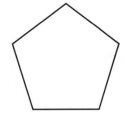

_____ is not shaded.

8. $\frac{1}{3}$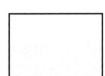

_____ is not shaded.

Name _____ Date _____

Test

Make 1 using the same unit fraction. Write how many more Fraction Bars you will need to make 1.

9.

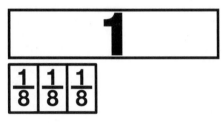

_____ more $\frac{1}{8}$ Fraction Bars

10.

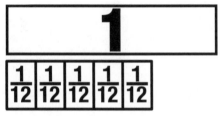

_____ more $\frac{1}{12}$ Fraction Bars

Write each total using both cent notation and dollar notation.

9.

_____ _____

10.

_____ _____

11.

_____ _____

12.

_____ _____

Chapter 6

Name _____ Date _____

Test

Write each fraction in decimal form.

1. $\frac{2}{5}$ _____

2. $\frac{3}{4}$ _____

3. $\frac{1}{3}$ _____

4. $\frac{4}{10}$ _____

Write each mixed number as an improper fraction.

5. $3\frac{1}{2}$ _____

6. $1\frac{2}{3}$ _____

7. $5\frac{4}{5}$ _____

8. $2\frac{6}{7}$ _____

Complete the chart below so that the fraction and decimal equal each other.

	Fraction	Decimal
9.	$-\frac{1}{8}$	
10.		-2.0
11.	$-\frac{4}{5}$	
12.		-0.4

Name _____ **Date** _____

Test

Label each missing number on each number line.

13.

0.1 0.2 0.3 ____ 0.5 0.6 0.7 0.8 ____ 1.0 ____

14.

____ $-\frac{1}{4}$ ____ $\frac{1}{4}$ ____ $\frac{3}{4}$ 1 $1\frac{1}{4}$ $1\frac{2}{4}$ $1\frac{3}{4}$ 2

15.

$\frac{1}{10}$ $\frac{2}{10}$ ____ $\frac{4}{10}$ $\frac{5}{10}$ $\frac{6}{10}$ ____ $\frac{8}{10}$ ____ $\frac{10}{10}$ $1\frac{1}{10}$

16.

-1.0 ____ -0.6 ____ -0.2 0 0.2 ____ 0.6 0.8 1.0

Estimate where each decimal is on the number line. Label each with a point and the decimal.

17. 0.1 and 0.6

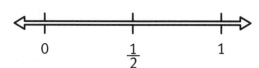

0   $\frac{1}{2}$ 1

19. -0.5 and -0.4

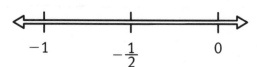

-1 $-\frac{1}{2}$ 0

18. 0.25 and 1.0

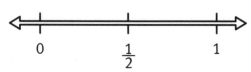

0   $\frac{1}{2}$ 1

20. -0.75 and -0.1

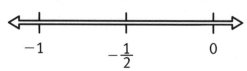

-1 $-\frac{1}{2}$ 0

Chapter 7

Name _____ Date _____

Test

Circle the prime numbers.

1. 7, 8, 9

2. 10, 11, 12

3. 22, 23, 24

4. 39, 40, 41

Construct composite numbers. Write your answer.

5. Construct two composite numbers that are evenly divisible by 5 and 4.

Write the following numbers as the product of their prime factors.

6. 44 _____

7. 32 _____

8. 28 _____

9. 22 _____

10. 9 _____

11. 50 _____

Test

Write the base with its exponent.

12. $6 \times 6 \times 6 \times 6 =$ _____

13. $5 \times 5 \times 5 \times 5 \times 5 =$ _____

14. $10 \times 10 =$ _____

15. $8 \times 8 \times 8 =$ _____

16. $9 \times 9 \times 9 \times 9 =$ _____

17. $20 \times 20 \times 20 \times 20 \times 20 =$ _____

18. $7 \times 7 \times 7 \times 7 \times 7 \times 7 \times 7 =$ _____

19. $12 \times 12 \times 12 =$ _____

Write each number in standard form.

20. $7 \times 10^3 + 8 \times 10^2 + 3 \times 10^1 + 8 \times 10^0 =$ _____

21. $1 \times 10^5 + 0 \times 10^4 + 5 \times 10^3 + 2 \times 10^2 + 1 \times 10^1 + 0 \times 10^0 =$

_____

22. $3 \times 10^3 + 4 \times 10^2 + 3 \times 10^1 + 0 \times 10^0 =$ _____

23. $3 \times 10^2 + 8 \times 10^1 + 9 \times 10^0 + 1 \times 10^{-1} + 0 \times 10^{-2} + 5 \times 10^{-3} =$

_____

24. $7 \times 10^4 + 3 \times 10^3 + 1 \times 10^2 + 6 \times 10^1 + 0 \times 10^0 =$ _____

Chapter 8

Name _____ Date _____

Test

Rewrite the fractions in each pair so they have the same denominator, and write a true statement using <, >, or =.

1. $\frac{2}{6}$ $\frac{4}{11}$ _____ _____ _____

2. $\frac{3}{4}$ $\frac{4}{5}$ _____ _____ _____

3. $\frac{1}{8}$ $\frac{2}{10}$ _____ _____ _____

4. $\frac{4}{5}$ $\frac{7}{10}$ _____ _____ _____

5. $\frac{1}{4}$ $\frac{2}{8}$ _____ _____ _____

6. $\frac{1}{6}$ $\frac{1}{10}$ _____ _____ _____

7. $\frac{5}{6}$ $\frac{9}{11}$ _____ _____ _____

8. $\frac{8}{12}$ $\frac{2}{3}$ _____ _____ _____

Write each as a part of 100 using fraction and decimal form.

9. 23 pennies   _____ _____

10. 7 pennies   _____ _____

11. 58 pennies   _____ _____

12. 100 pennies   _____ _____

Name _____ Date _____

Test

Complete the chart below so the percent, fraction, and decimal in each row equal one other.

Percent	Fraction	Decimal
	$\frac{1}{10}$	
		0.3
80%		
		0.0

Write the answer to each question. You can draw models to help find percentages.

25% of 12

13. 25% is the same as _____.

14. _____ of 12 is _____.

15. 25% of 12 is _____.

60% of 20

16. 60% is the same as _____.

17. _____ of 20 is _____.

18. 60% of 20 is _____.

Estimate where each number is on the number line. Label each with a point and the given value.

19. 0.1 and 0.8

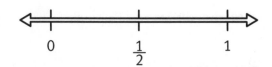

20. 40% and 60%

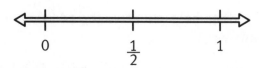

Name a percentage between the given values. Record it on the number line.

21.

_____

22.

_____

Name _____ Date _____

Test

Use the number line to add or subtract numbers. Circle the starting number. Use arrows to show the amount added or subtracted. Put a square around the answer.

1. $-3 - 2 =$ _____

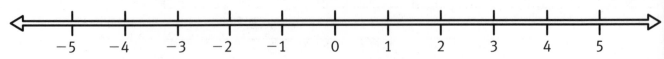

2. $5 + -1 =$ _____

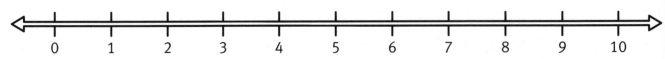

Use the number line to help you solve each problem.

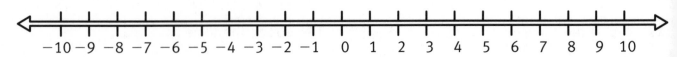

3. $12 + 3 =$ _____

4. $-2 + -8 =$ _____

5. $5 - 6 =$ _____

6. $-1 - (-3) =$ _____

Use grids to find each sum or difference.

7. $\frac{1}{10} + \frac{5}{10} =$ _____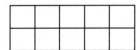

8. $\frac{5}{6} - \frac{4}{6} =$ _____

Use grids to help you solve the addition or subtraction problems. Write your answers.

9. Montega had 10 colored pencils in his box. What fraction of the box of pencils is left after Montega gives his sister 4 pencils?

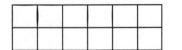   _____ of a box

10. A school district recycled $\frac{1}{8}$ pound of aluminum yesterday and $\frac{1}{4}$ pound of aluminum today. How much more aluminum did the district recycle today than yesterday? Draw your own grid.

_____ pound

Name _____ Date _____

Test

Use the grid to add or subtract the decimals.

11. $0.23 - 0.16 =$ _____

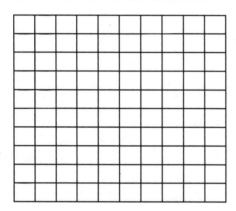

12. $0.31 + 0.47 =$ _____

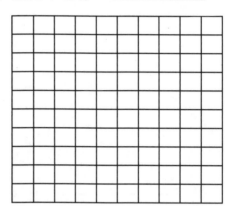

Add or subtract the decimals.

13. $0.42 + 0.33 =$ _____

14. $0.8 - 0.12 =$ _____

15. $0.81 + 0.03 =$ _____

16. $0.07 + 0.03 =$ _____

17. $0.34 + -0.22 =$ _____

18. $-0.64 - (-0.81) =$ _____

Solve the equations.

19. $12^3 =$ _____

20. $1^{20} =$ _____

21. $250^1 =$ _____

22. $58^0 =$ _____

23. $3^4 =$ _____

24. $7^2 =$ _____

25. $5^3 =$ _____

26. $10^4 =$ _____

Name _____ Date _____

Test

Multiply or divide.

1. $17 \times -4 =$ _____

2. $-25 \times -5 =$ _____

3. $40 \div -8 =$ _____

4. $28 \times -64 =$ _____

5. $-46 \times -71 =$ _____

6. $120 \div -6 =$ _____

7. $62 \div 31 =$ _____

8. $-84 \div 12 =$ _____

Reduce each fraction by dividing the numerator and the denominator by a common factor. Write your answer.

9. $\frac{32}{8} =$ _____

The common factor I divided the numerator and denominator by was

_____.

10. $\frac{42}{12} =$ _____

The common factor I divided the numerator and denominator by was

_____.

Multiply or divide the fractions. Write each answer as a reduced fraction.

11. $5 \times \frac{1}{10} =$ _____

12. $\frac{1}{20} \div \frac{7}{4} =$ _____

13. $-12 \times -\frac{3}{4} =$ _____

14. $\frac{2}{5} \times 30 =$ _____

15. $\frac{4}{5} \times \frac{5}{4} =$ _____

16. $-7 \div \frac{1}{14} =$ _____

17. $\frac{1}{3} \times \frac{3}{8} =$ _____

18. $24 \times \frac{1}{240} =$ _____

Test

Use the grid to multiply the decimals.

19. $0.5 \times 0.3 =$ _____

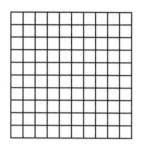

20. $0.6 \times 0.9 =$ _____

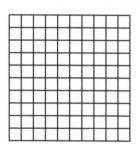

21. $0.1 \times 0.4 =$ _____

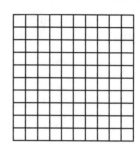

22. $0.8 \times 0.3 =$ _____

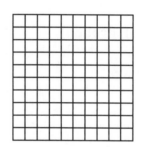

Multiply the decimals. Show your work, and write your answer.

23. A cheetah is running 42.1 miles per hour. How far will the cheetah

travel over a period of 1.75 hours? _____ miles

24. Chet is making fruit salad for his classmates. He can peel 3.5 oranges per minute. How many oranges can he peel over a period of 12 minutes?

_____ oranges

Multiply the decimals by the powers of ten. Write the product.

25. $10 \times 0.345 =$ _____

26. $0.01 \times 76.5 =$ _____

Multiply or divide the expressions as indicated. Write the result.

27. $0.016 \times 0.25 =$ _____

28. $0.512\overline{)5.632} =$ _____

29. $8.03\overline{)56.21} =$ _____

30. $-9 \times 2.28 =$ _____

31. $-9.2 \times -0.058 =$ _____

32. $0.37 \div 8.51 =$ _____

Chapter 11

Name _____ Date _____

Test

Complete the tables below so the percent, fraction, and decimal in each row equal each other.

	Percent	Fraction
1.	3%	
2.		$\frac{9}{20}$
3.	70%	
4.		$\frac{16}{25}$

	Percent	Fraction
5.	52%	
6.		0.06
7.		0.3
8.		

	Percent	Fraction
9.		0.8
10.	$\frac{1}{3}$	
11.		0.3
12.	$\frac{12}{25}$	

Shade a model for each fraction.

13. $\frac{1}{6}$ = $\frac{2}{12}$

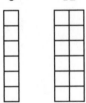

14. Create another model to show a fraction that is equivalent to $\frac{1}{6}$ and

$\frac{2}{12}$. Write the fraction. _____

15. Create a model for the fraction $\frac{1}{5}$. Then create two more models to show fractions that are equivalent to $\frac{1}{5}$. Write the fractions below each model.

16. Create a model to show that the following fractions are equivalent.

$\frac{2}{3}$ = $\frac{4}{6}$

Solve for each unknown value *x*. Show your work. Write the solution.

17. $\frac{25}{100} = \frac{x}{500}$ _____

18. $\frac{12}{100} = \frac{72}{x}$ _____

19. $\frac{7}{100} = \frac{49}{x}$ _____

20. $\frac{60}{100} = \frac{x}{5}$ _____

Write a proportion for each problem. Write the solution to each problem.

21. There are 40 people in the school choir. Last night 10 percent of the

choir sang solos. How many people sang a solo last night? _____

22. In a local small lake, there are 2,000 fish. Of those fish, 60 fish are

bass. What percentage of the fish is bass? _____

Decide whether the percent would make you pay more or less in the following situations, or whether the percent stands alone. Solve the problems. Write *is added to the total, is subtracted from the total,* or *stands alone* and the solution.

23. Mrs. Rivera has taken out a $20,000 car loan at a bank that charges 6 percent interest at the end of every year. How much will Mrs. Rivera owe at the end of the year?

The percent _____. Therefore, the

amount owed will be _____.

24. Mali went to a restaurant for dinner. His service was very good, so he decided to tip 18 percent. If his bill was originally $25, how much money should he pay altogether?

The percent _____. Therefore, Mali should pay _____.

25. A literary magazine is having a fund-raiser. They get 70 percent of the money from selling subscriptions. If they sell $400 in subscriptions, how much money will they earn for their magazine?

The percent _____. Therefore, the magazine will earn _____.

26. A grocery store is having a 60 percent off sale on all of their fresh vegetables. If an eggplant usually costs $2, how much will it cost after the discount?

The percent _____. Therefore, the amount

paid will be _____.

Test

Write the following exponential expressions as multiplication problems.
Simplify them by writing them in standard form.

1. 5^2

In expanded form, this is _____.

Simplified into standard form, this is _____.

2. 1^9

In expanded form, this is

_____.

Simplified into standard form, this is _____.

3. 45^1

In expanded form, this is _____.

Simplified into standard form, this is _____.

4. 3^4

In expanded form, this is _____.

Simplified into standard form, this is _____.

5. 7^3

In expanded form, this is _____.

Simplified into standard form, this is _____.

6. 5^3

In expanded form, this is _____.

Simplified into standard form, this is _____.

7. 8^4

In expanded form, this is _____.

Simplified into standard form, this is _____.

8. 6^3

In expanded form, this is _____.

Simplified into standard form, this is _____.

9. 3^{-2}

In expanded form, this is _____.

Simplified into standard form, this is _____.

10. 2^{-6}

In expanded form, this is

_____.

Simplified into standard form, this is _____.

11. 5^{-1}

In expanded form, this is _____.

Simplified into standard form, this is _____.

12. 5^{-4}

In expanded form, this is

_____.

Simplified into standard form, this is _____.

13. 6^{-3}

In expanded form, this is _____.

Simplified into standard form, this is _____.

14. 4^{-2}

In expanded form, this is _____.

Simplified into standard form, this is _____.

Chapter 12

Name _____ Date _____

Test

Change the expressions with negative exponents into fractions with positive exponents in the denominator.

15. $12^{-2} =$ _____

18. $40^{-2} =$ _____

16. $5^{-1} =$ _____

19. $10^{-6} =$ _____

17. $17^{-3} =$ _____

20. $31^{-3} =$ _____

Rewrite the exponential expressions as repeated multiplication. Write the resulting exponent.

21. $3^5 \times 3^2 =$ _____ = _____

22. $9^5 \div 9^3 =$ _____ = _____

Multiply or divide the exponents. Write the result as an exponential expression.

23. $5^5 \times 5^6 =$ _____

26. $10^5 \times 10^7 =$ _____

24. $7^8 \div 7^8 =$ _____

27. $9^{11} \div 9^4 =$ _____

25. $4^1 \times 4^{15} =$ _____

28. $8^5 \div 8^{10} =$ _____

Chapter 13

Name _____ Date _____

Test

Evaluate the expression by completing the operations inside the parentheses first. Circle the correct answer.

1. $8 - (5 \times 3) =$ -7 9

2. $(11 - 5) \times 6 =$ -19 36

Draw parentheses around the operation that should be performed first in order to get the correct answer.

3. $2 + 7 \times 4 = 30$

4. $15 - 2 + 8 = 5$

Write the solution to the expressions.

5. $1 + (9 \times 3) =$ _____

6. $7 \times (4 + 6) =$ _____

Use the digits to reach the target number. Write each digit one time in the correct space.

7.   Target number: 23

Digits to combine: 1, 3, 5, 8

Equation: (_____ × _____) + (_____ × _____)

8.   Target number: 20

Digits to combine: 0, 2, 3, 4

Equation: (_____ + _____) × (_____ + _____)

Use the operations to reach the target number. Write each operation one time in the correct space.

9.   Target number: 4

Operations to combine: ÷, +, +

Equation: (14 _____ 2) _____ (1 _____ 3)

10.   Target number: -18

Operations to combine: ×, ×, −

Equation: (2 _____ 1) _____ (5 _____ 4)

Test

Write the first operation you would perform in each equation.

11. $1^{-2} + 3$

The first operation is _____.

13. $2 + 4 \div 2$

The first operation is _____.

12. $10 + (5 + 1) \div 2$

The first operation is _____.

14. $2 \times 4 \times 15$

The first operation is _____.

Evaluate the following expressions. Write the solution.

15. $11 \times (1 + 4) + 3^2 =$ _____

17. $15 \times 2^2 + (1 + 5) =$ _____

16. $1^3 - 5 \div (6 - 1) =$ _____

18. $4^2 + (2 \times 3)^2 \times 3 =$ _____

Write the expressions and solutions that are described.

19. Namid's friend in 10 years will be twice the current age of Namid. If Namid's friend is 16, then how old is Namid?

_____

Namid is _____ years old.

20. If you add 40 to a number, you get $\frac{1}{2}$ of 100. What is the number?

_____

The unknown number is _____.

Write the solutions to the expressions that are described. It may be useful to use counters to model the expressions.

21. If you subtract 11 from a number, you get 5 times 5. What is the number?

_____

22. The visiting basketball team had 40 points more than $\frac{1}{2}$ the score of the home team. If the home team had 70 points, then what score did the visiting team have?

_____

Chapter 14

Name _____ Date _____

Test

Write the evaluations of the expressions with the variables as defined.

1. Expression: $0.2a$

Variable: $a = 20$

Evaluation: _____

2. Expression: $-3b$

Variable: $b = 5$

Evaluation: _____

Write the expression that would result in the outputs desired.

3.

Input	Output
0	0
1	−4
2	−8
3	−12

Expression: _____

4.

Input	Output
0	3
1	4
2	5
3	6

Expression: _____

Write the solution for each variable. It might be useful to use counters to model the situations.

5. $11 + a = 2$

_____

6. $7b = 42$

_____

Complete the table by writing the missing input or output for the expression $7x$.

	Input	Output
7.	$x = -1$	
8.		70

Test

Draw a number line to represent the quantities described by the expressions.

9. 7 plus a number

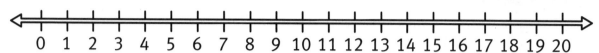

10. 16 divided by a number

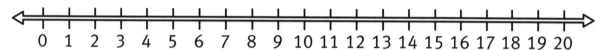

Write the expression that is described.

11. The product of 7 and a number

_____

12. The sum of 1 and a number

_____

Draw a number line to represent the quantities described in the expressions. **Write the expressions.**

13. A number is less than 15.

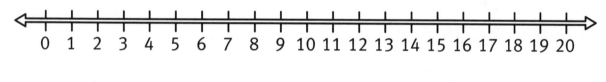

_____

14. A number is more than or equal to 6.

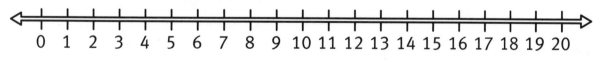

_____

Write the equation that is described.

15. A number y is equal to the product of 7 and x.

_____

16. A number y is equal to x to the third power.

_____

Test

Write the *y*-value in the table for the equation $y = 5 + 3x$.

	x	y
1.	−7	
2.	1	
3.	2	
4.		35

Create a table to represent some solutions for the following two-variable equations.

5. $y = x - 2$

x	y

6. $y = -2x$

x	y

If the first scale is balanced, will the second scale be balanced? Write *yes* or *no*.

7.

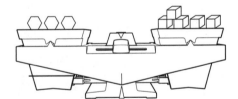

_____

8.

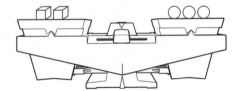

_____

Write the number that makes the statement true.

9. $q = r$; $q + 10 = r +$ _____

10. $a = b$; $a - 7 = b -$ _____

Name _____ Date _____

Test

If the first scale is balanced, will the second scale be balanced? Write *yes* or *no*.

11.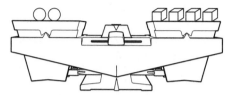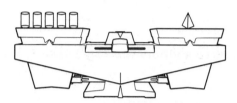

_____

12.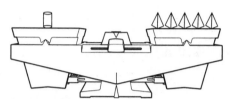

_____

Write the number that makes the statement true.

13. $a = b$; $0.8a = $ _____ b **14.** $d = t$; $2d = ($ _____ $- 11)t$

Write the new equations that result after performing the indicated operations to both sides of the equations.

15. $x - 8 = 30$; add 8 **16.** $2x = 15$; divide by 2

_____ _____

Select an operation to help you isolate the variable, and then perform that operation on both sides. Write the resulting equation.

17. $x - 4 = 7$ **18.** $-12x = 48$

The operation I performed is: The operation I performed is:

_____. _____.

The resulting equation is: The resulting equation is:

_____. _____.

Isolate the variable. Write the solution.

19. $x + 2 = 14$ _____ **20.** $3x = 45$ _____

Name _____ Date _____

Test

Write which property you can use to simplify the expressions.

1. $-2 \times \frac{1}{-2}$

_____

3. $(2 \times 11) \times 5$

_____

2. $15 + 0$

_____

4. $12 + (6 + 8)$

_____

Apply the properties to simplify the expressions. Write the resulting expressions.

5. $7 \times (10 + 8)$; distributive property

_____

7. $13 + (-13)$; inverse property of addition

_____

6. $0 + 4$; identity property of addition

_____

8. $\frac{1}{8} \times 8$; inverse property of multiplication

_____

Write the solutions to the two-step equations. Verify your solutions by substituting for *x*. Show your work.

9. $x \times 4 - 2 = 30$ _____

13. $\_\_ \times 4 - 2 = 30$ _____

10. $3x \div 15 = 1$ _____

14. $3\_\_ \div 15 = 1$ _____

11. $-x - 3 = 20$ _____

15. $-\_\_ - 3 = 20$ _____

12. $x + 22 = 45$ _____

16. $\_\_ + 22 = 45$ _____

Name _____ Date _____

Test

Write the solutions to the two-step equations. Verify your solutions by substituting for *x*. Show your work.

17. $7x - 3 < 39$ _____

18. $x \div 3 + 2 < 10$ _____

19. $-x + 2 = 16$ _____

20. $6x + 4 \geq 10$ _____

21. $7\underline{} - 3 < 39$ _____

22. $\underline{} \div 3 + 2 < 10$ _____

23. $-\underline{} + 2 = 16$ _____

24. $6\underline{} + 4 \geq 10$ _____

Write the rate for each of the problems.

25. A certain boat can travel 100 miles in 4 hours. How many miles per hour can it travel?

_____

26. Marquis earned 225 dollars last week. If he worked for 15 hours, how much does he earn per hour?

_____

Answer the following questions.

A certain person eats a total of 21 ounces of vegetables for dinner and 49 ounces of vegetables for lunch in 7 days. How many ounces of vegetables does he eat altogether in 5 days?

27. What rates do you need to find?

_____

28. What are these rates?

_____

29. How do you use these rates to find out how many ounces of vegetables he ate for dinner and how many ounces of vegetables he ate for lunch over 5 days?

_____

30. How many ounces of vegetables does he eat altogether in 5 days?

_____

Name _____ Date _____

Test

Give the coordinates of the point indicated by the letter. Be sure to use correct notation.

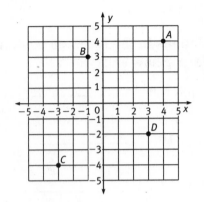

1. A _____

2. B _____

3. C _____

4. D _____

Describe how to get from the first point listed to the second point listed. Use a coordinate grid to help you answer the questions. Be sure to describe left-right movement first and then up-down movement.

5. How would you move from $(-8, 1)$ to $(3, 4)$?

_____

6. How would you move from $(2, 5)$ to $(-5, -2)$?

_____

7. How would you move from $(0, 11)$ to $(0, 9)$?

_____

8. How would you move from $(-5, 4)$ to $(-8, 6)$?

_____

Write the quadrant in which each point is located. You can use a coordinate grid to help you identify the location. If the point is not in a quadrant, list the axis or origin on which it lies.

9. $(-8, -10)$ _____

10. $(5, 6)$ _____

11. $(-10, 8)$ _____

12. $(1, 0)$ _____

13. $(-7, 5)$ _____

14. $(0, 1)$ _____

Plot the given point on the coordinate grid.

15. $(4, 1)$

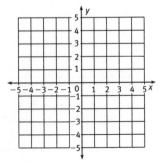

16. $(2, -5)$

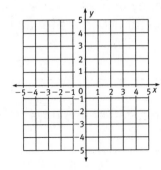

17. $(-5, -4)$

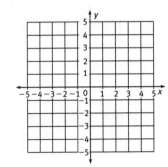

18. $(-1, -5)$

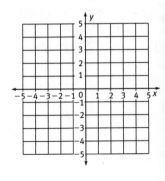

Name _____ Date _____

Test

Plot the given points on the coordinate grid. Use a ruler to draw a line that goes through the two points you have plotted on the grid.

19. (−3, −4) and (−1, 2)

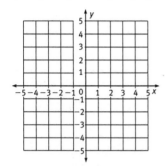

20. (−2, −2) and (−3, 0)

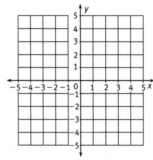

21. (3, −3) and (5, 5)

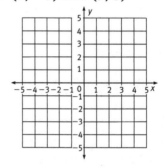

22. (−3, 2) and (−3, 4)

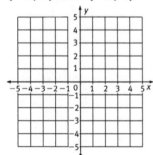

23. (−3, 2) and (3, 5)

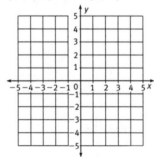

24. (4, 4), (3, 3), and (2, 2)

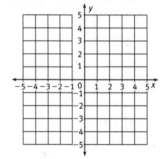

Follow the instructions to create shapes, and then identify the shapes you created.

25. Graph these points on the coordinate plane: A (−2, 2); B (2, 2); C (0, 4). Then connect the points in alphabetical order. Connect points A and C. Identify the shape you have made.

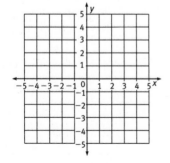

The shape is a(n) _____.

26. Graph these points on the coordinate plane: A (0, 0); B (1, 3); C (2, 4); D (3, 3); E (4, 0). Connect points A, B, C, D, and E in alphabetical order. Now connect point B to point D. Identify the shape you have made.

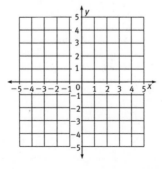

The shape is a(n) _____.

Test

Determine the length of the line segments that connect these pairs of coordinates. Show your work.

1. (2, 7) and (−8, 7)

_____

2. (10, 1) and (9, 1)

_____

3. (−9, −3) and (−9, −3)

_____

4. (2, 4) and (−8, 4)

_____

5. (2, −8) and (2, 7)

_____

6. (−5, 2) and (1, 2)

_____

Determine the length of the line segments that connect these pairs of coordinates. Show your work.

7. (9, 10) and (9, −9)

_____

8. (5, 8) and (5, 7)

_____

9. (−9, −3) and (−9, −1)

_____

10. (5, −2) and (5, 9)

_____

11. (0, −4) and (0, −2)

_____

12. (7, 4) and (7, −7)

_____

Name _____ Date _____

Test

Use the Pythagorean theorem to find the length of the hypotenuse.
Show your work below.

13. $a = 5, b = 12, c =$ _____

15. $a = 15, b = 36, c =$ _____

14. $a = 12, b = 16, c =$ _____

16. $a = 6, b = 8, c =$ _____

Consider these groups of three numbers, and determine whether
they are the sides in a right triangle. Explain your answers. You will
use the Pythagorean theorem. In each case, the hypotenuse will be
identified by c.

17. 4, 6, $c = 10$

_____

_____

20. 4, 5, $c = 6$

_____

_____

18. 3, 9, $c = 11$

_____

_____

21. 7, 7, $c = 14$

_____

_____

19. 6, 8, $c = 10$

_____

_____

22. 12, 5, $c = 13$

_____

_____

Answer the questions about the graphs as they appear. Each graph is identified by a capital letter.

A.

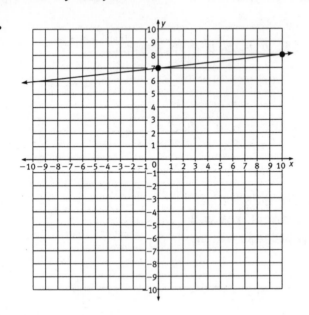

B.

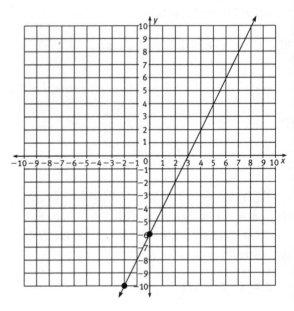

1. The graphs in order from greatest slope to

least slope are _____.

2. The slope of the line in Graph A is _____.

3. The slope of the line in Graph B is _____.

4. The graphs in order from least slope to

greatest slope are _____.

Use the formula for slope given two points, and find the slope of the line through each of these pairs of points.

5. $(2, 2)$ and $(-2, -10)$ _____

6. $(-3, 6)$ and $(-9, 1)$ _____

7. $(-7, 5)$ and $(-9, -1)$ _____

8. $(-2, -9)$ and $(-6, -6)$ _____

9. $(-8, 9)$ and $(-3, -5)$ _____

10. $(3, -5)$ and $(1, 9)$ _____

11. $(-2, 5)$ and $(1, 6)$ _____

12. $(-4, -5)$ and $(-9, -6)$ _____

Use the graph and the equation of $y = 3x$ to answer the following questions.

13. When $x = 2$, what is the value of y? _____

14. When $y = 4$, what is the value of x? _____

15. When $y = 0$, what is the value of x? _____

16. When $x = -4$, what is the value of y? _____

17. When $y = -6$, what is the value of x? _____

18. When $x = 1$, what is the value of y? _____

19. When $y = -12$, what is the value of x? _____

20. When $y = -1$, what is the value of x? _____

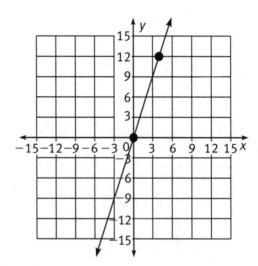

Graph the relationship between the perimeter of an equilateral triangle and the length of its sides using the formula $P = 3s$.

21. Write the formula as an equation involving x and y.

_____

22. Write three ordered pairs of numbers that can help you graph this line. Remember to keep the x-values fairly small so that the graph does not get too large.

_____

23. Plot the three points that you picked, and draw the line through those points.

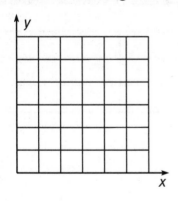

24. From the points you graphed on the coordinate grid, calculate the slope of the line (change in y-values/change in x-values or rise/run).

_____

Chapter 20

Name _____ Date _____

Test

Consider the equation $y = -x + 2$ to answer the following questions.

1. What is the *y*-intercept of this line?

_____

2. Write the coordinates of the *y*-intercept.

_____

3. What is the slope of this line?

_____

4. From the *y*-intercept, the slope tells you to move up _____ unit(s) and then to the _____ unit(s) to find another point on this line. The coordinates of this point are _____.

5. From the y-intercept, you could also move down _____ unit(s) and then back to the _____ unit(s) to find a third point on this line. The coordinates of this point are _____.

Complete the following chart that shows the payment information for summer part-time jobs. Find the hourly rate (money per hour) that each student earned.

6. Wage Comparison

Student Name	Amount of Paycheck	Number of Hours Worked	Rate per Hour
Diego	$106.50	15	
Susanna	$174.00	24	
Eldora	$122.40	18	
Manuel	$140.00	20	

7. Which student had the highest rate per hour? _____

8. Which student earned the least per hour? _____

Name _____ Date _____

Test

Use the following information to answer the questions.

A contractor needs to frame the houses in a small housing development. It will take 35 days for his crew of 6 framers to complete this project.

9. How many person-days are needed for

this project? _____

10. At this rate, how long would it take
1 framer to complete the job alone?

_____

11. At this rate, how much of the project does
each framer complete in 1 day?

_____

12. How many days would it take if the
contractor hired 5 framers?

_____

13. How many days would it take if the
contractor hired 10 framers?

_____

14. How many days would it take if the
contractor hired 12 framers?

_____

15. How many framers would the contractor
need to hire to complete this job in

5 days? _____

16. How many framers would the contractor
need to hire to complete this job in

15 days? _____

Find the error in each solution. Then give the correct answer.

17. Change 2 rotations per second to rotations per day.

$$\frac{2 \text{ rot}}{1 \text{ sec}} \times \frac{60 \text{ sec}}{1 \text{ min}} \times \frac{1 \text{ min}}{60 \text{ hrs}} \times \frac{24 \text{ hrs}}{1 \text{ day}} =$$

$$\frac{2}{1} \times \frac{60}{1} \times \frac{1}{60} \times \frac{24}{1} = 48 \text{ rotations per day}$$

_____

The correct answer is _____.

Name _____ Date _____

Test

Fill in the blanks with *opposite* or *reciprocal* to make true statements.

1. 5 is the _____ of -5.

2. 5 is the _____ of $\frac{1}{5}$.

3. -10 is the _____ of $-\frac{1}{10}$.

4. $\frac{5}{3}$ is the _____ of $\frac{3}{5}$.

5. $-\frac{1}{6}$ is the _____ of -6.

Complete the following chart.

6.

Number	Opposite	Reciprocal
5	-5	$\frac{1}{5}$
-12		
0.4		
$-\frac{1}{8}$		
$-\frac{1}{2}$		
$\frac{1}{2}$		

Simplify each expression by correctly using the distributive property.

7. $-4(x + 3)$ _____

8. $(-8x - 40) \div 8$ _____

9. $3(x - 2)$ _____

10. $3(1 - 2x)$ _____

11. $(4x - 8) \div 2$ _____

12. $5(-8x + 6)$ _____

13. $(-15x + 30) \div -5$ _____

14. $(14x - 21) \div 7$ _____

15. $-4(-2 - x)$ _____

16. $-(3x - 1)$ _____

Name _____ Date _____

The power equation $5^4 = 625$ and the root equation $\sqrt[4]{625} = 5$ are inverses of each other. Write the inverse of each given equation.

17. If $\sqrt{81} = 9$, then _____.

20. If $\sqrt[3]{64} = 4$, then _____.

18. If $5^3 = 125$, then _____.

21. If $17^2 = 289$, then _____.

19. If $14^2 = 196$, then _____.

22. If $6^4 = 1{,}296$, then _____.

Simplify each variable expression.

23. $\sqrt{4x^8} =$ _____

25. $\sqrt[4]{81x^{40}} =$ _____

24. $\sqrt[5]{32x^{20}} =$ _____

26. $\sqrt[15]{x^{30}} =$ _____

Write the letter of the power equation in the right-hand column that matches its inverse power equation from the left-hand column.

27. $81^{\frac{3}{4}} = 27$ _____

A. $27^{\frac{4}{3}} = 81$

28. $15^2 = 225$ _____

B. $125^{\frac{2}{3}} = 25$

29. $64^{\frac{1}{2}} = 8$ _____

C. $\sqrt{49} = 7$

30. $25^{\frac{3}{2}} = 125$ _____

D. $\sqrt{225} = 15$

31. $7^2 = 49$ _____

E. $1^5 = 1$

32. $1^{\frac{1}{5}} = 1$ _____

F. $8^2 = 64$

Find the missing leg or hypotenuse. If your answers are not perfect squares, leave them in square-root form. The hypotenuse is *c*.

1. $a = 11$; $c = 16$; find *b*.

   $b =$ _____

2. $a = 8$; $b = \sqrt{36}$; find *c*.

   $c =$ _____

3. $b = 2$; $c = 5$; find *a*.

   $a =$ _____

4. $b = \sqrt{25}$; $c = \sqrt{169}$; find *a*.

   $a =$ _____

5. $a = \sqrt{2}$; $c = \sqrt{3}$; find *b*.

   $b =$ _____

6. $a = 4$; $b = 7$; find *c*.

   $c =$ _____

Write whether the following statements are true or false. If the statement is false, write the correct answer.

7. $9^0 = 90$

   _____

8. $5^{-2} = -25$

   _____

9. $-10^{-6} = \dfrac{1}{10^6}$

   _____

10. $-2^5 = -32$

    _____

11. $2^{-4} = \dfrac{1}{16}$

    _____

12. $-3^{-1} = -\dfrac{1}{3}$

    _____

13. $4^{-10} = -104$

    _____

14. $11^0 = 1$

    _____

15. $5^1 = 1$

    _____

16. $1^{-15} = 1$

    _____

Simplify. Assume variables do not equal 0.

17. $(7y)^{-2} =$ _____

18. $(2x)^5 =$ _____

19. $(-3)^3 =$ _____

20. $(8x^3)^{\frac{1}{3}} =$ _____

21. $(-4x^4)^0 =$ _____

22. $(3x^4)^1 =$ _____

Write whether the following statements are true or false. If the statement is false, write the correct answer. Leave the answers in exponential form.

23. $\frac{5^{11}}{5^5} = 5^3$

_____

24. $(1^3)^{-1} = 1$

_____

25. $(3x)^3 = 9x^3$

_____

26. $(\frac{1}{4})^2 = 16$

_____

27. $(3x)^2 \times (3x)^2 = (3x)^4$

_____

28. $11^3 \times 11^8 = 11^{11}$

_____

29. $(x)^8 \times (x)^{11} = (x)^{88}$

_____

30. $\frac{(2x)^6}{(2x)^2} = (2x)^3$

_____

Simplify the following expressions. Assume variables do not equal 0.

31. $(\frac{4}{x})^2 =$ _____

32. $(5x^2)^{-1} =$ _____

33. $(12xy)^0 =$ _____

34. $(-x)^2 =$ _____

35. $(3x)^2 \times (3x)^4 =$ _____

36. $\frac{(2x)^3}{(2x)^5} =$ _____

Simplify using the 5 Basic Rules, PEMDAS, the distributive property, and the rules for signed numbers. Assume variables do not equal 0.

37. $(-1x)^5 =$ _____

38. $(2x^2)^2 =$ _____

39. $-2(2x - 1) =$ _____

40. $10x^3 - 10x^3 =$ _____

41. $\frac{x^7}{x^9} =$ _____

42. $-x^4 + 4x^3 =$ _____

43. $x^9 - x^9 + 2x^9 =$ _____

44. $(-11x^0)^4 =$ _____

45. $5x^2(-1 + 2x - x^2) =$ _____

46. $(-2)^{-3} \times (-2)^5 =$ _____

47. $\frac{-x^2}{6x^2} =$ _____

48. $(-11x)(5x) =$ _____

49. $\left(\frac{45x}{28}\right)^0 =$ _____

50. $-6x^2(-2x - 3) =$ _____

51. $\frac{2x^4}{-8x^5} =$ _____

52. $-x(4x^2 - 5x) =$ _____

53. $-1^9 =$ _____

54. $\frac{10x^2}{5x^7} =$ _____

Chapter 23

Name _____ Date _____

Test

Write the letter that matches the expression in Column 1 with its correct simplification in Column 2.

1. $10x + 12x^6 - 7x + x^4$ _____
2. $(3x^2)(-x^4)$ _____
3. $(-x^6)(-x^{-3})$ _____
4. $-2x^5 - 10x^5$ _____
5. $\dfrac{x^7}{-3x^{-2}}$ _____
6. $2x^2 + 3x^3 - 10x^2 + 2x^3 + 3x^2$ _____
7. $2x^3 + 3x - 4 + x - 8$ _____
8. $\dfrac{7x^5}{x^5}$ _____
9. $11x^{-4} + -2x^{-4}$ _____
10. $-4x^2 - 9x^2$ _____

A. $2x^3 + 4x - 12$
B. $-12x^5$
C. x^3
D. $-5x^2 + 5x^3$
E. $12x^6 + x^4 + 3x$
F. $-13x^2$
G. $-\dfrac{x^9}{3}$
H. $9x^{-4}$
I. $-3x^6$
J. 7

Solve. Write each step, and be prepared to explain the property, rule, or process that you used.

11. $a - 2 = 11$ _____
12. $2 = k + 4$ _____
13. $\dfrac{r}{-7} = 5$ _____
14. $-4y = 20$ _____
15. $-7c - 2 = 54$ _____
16. $-8m + 1 = \dfrac{19}{3}$ _____
17. $-6r - 5 = -17$ _____
18. $3c - 10 = 5$ _____
19. $2m + 2m + -2m = 10$ _____

20. $5 = 4k - k$ _____
21. $-y = 5$ _____
22. $8r - 4r + 4r = 0$ _____
23. $-2y - 4 = 2$ _____
24. $-15n + 2 = 32$ _____
25. $-a - 6 = 7$ _____
26. $4x - 2x - x = 15$ _____
27. $-10y + 3 = -7$ _____
28. $2a - 1.2 = 0$ _____

There is an error in each simplification. Explain the error, and simplify the expression correctly.

29. $-4c^2(c + c^2) = -4c^3 - 4c^2$

The error is _____

_____.

The correct simplification is _____.

30. $-11(c + 4) = -11c + 44$

The error is _____

_____.

The correct simplification is _____.

31. $-2c - (5c) + 4c^4 = -3c + 4c^2$

The error is _____.

_____.

The correct simplification is _____.

32. $5c + (-5c + 15c^2) = 5c + (10c^2)$

The error is _____.

_____.

The correct simplification is _____.

Solve each equation. Show your steps.

33. $5a - 2 + 4a = 25$

$a =$ _____

34. $-5(2a - 4) = -10 - 20$

$a =$ _____

35. $4(5a - 1) = 56$

$a =$ _____

36. $2 + 5(-a - 6) = -13$

$a =$ _____

37. $(11a - 5) + (a - 4) = -21$

$a =$ _____

38. $\frac{1}{4}(40a - 40) = -5 + 5$

$a =$ _____

Solve each equation, and prove your answer. Show your work.

39. $-42 = 5a + 2 - a + 4$

$a =$ _____

Check statement: _____

40. $5 - 5(1 + 9a) = 110 + 25$

$a =$ _____

Check statement: _____

41. $5a - (2a - 3) + 4a = 17(5 - 4)$

$a =$ _____

Check statement: _____

42. $(a - 10) = 2^4 - 11$

$a =$ _____

Check statement: _____

Translate each sentence into an equation, and solve. To prove that your answer is correct, write each problem with the answer in place. If it makes a true statement, write *true*.

1. Two less than four times a number is 22.

   Equation: _____ Answer: _____

   Proof: _____

2. Six times the difference of 5 and a number is 18.

   Equation: _____ Answer: _____

   Proof: _____

3. The quotient of a number and 4 is 20.

   Equation: _____ Answer: _____

   Proof: _____

Solve each inequality, and graph the solution.

4. $y - 3 < 5$ The solution is _____.

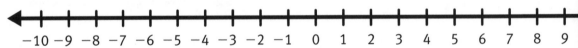

5. $10y + 7 \geq 17$ The solution is _____.

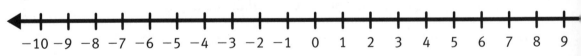

6. $3(y - 1) - 11 < 7$ The solution is _____.

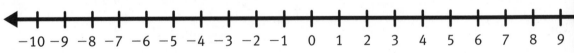

Use an inequality to answer the question. You will need to use the four-step process to solve this problem.

The juice bar has two pricing structures. A souvenir mug costs $5 plus $0.50 each time you purchase a juice. A regular mug costs $1.75 each time you purchase a juice. How many juices would someone have to order before the souvenir mug costs less than purchasing a regular mug each time?

7. Identify the variable.

_____

8. Express all unknown quantities in terms of the variable.

_____

_____

9. Set up the inequality.

_____

10. Solve, and conclude.

_____

_____

_____

Match the terms on the left with their correct definitions on the right.

11. difference

12. product

13. opposite
14. sum
15. quotient

16. inequality

17. reciprocal

18. root

19. equation
20. expression

A. A quantity that, when multiplied by itself a specified number of times, produces a given quantity

B. A number sentence stating that two quantities are not equal; relation symbols for inequalities include < (less than) and > (greater than)

C. The result of adding two or more numbers
D. The answer to a division problem
E. The remainder left after subtracting one quantity from another

F. A group of mathematical symbols (numbers, operations signs, variables, grouping symbols) that represents a number

G. A number that can be added to the original number and result in zero

H. The result of multiplying two or more numbers

I. Two quantities whose product is one

J. A mathematical statement showing that one quantity or expression is equal to another quantity or expression

Name _____ Date _____

Unit Test

For each pair, write the place value where the numbers differ.

1. 5,185   5,285   _____

2. 527   547   _____

3. 16,483   16,484   _____

4. 6,417   9,417   _____

Find the sum. Write your answer.

5.   5317
   + 452

7.   1414
   + 56

6.   1356
   + 1742

8.   508
   + 808

Find each difference. Write your answer. Use addition to check your answer.

9.   2236   Check: 67
   − 67   + _____
                     2236

12.   2645   Check:
   − 751   + _____

10.   897   Check: 43
   − 43   + _____
                    897

13.   2449   Check:
   − 489   + _____

11.   679   Check:
   − 78   + _____

14.   3316   Check:
   − 624   + _____

Name _____ Date _____

Unit Test

Break the second factor of each problem into a sum. Find each product.
Write your answer.

15. 8×28

(8 × _____) + (8 × _____) = _____

17. 4×27

(4 × _____) + (4 × _____) = _____

16. 12×52

(12 × _____) + (12 × _____) = _____

18. 3×67

(3 × _____) + (3 × _____) = _____

Use the traditional method to find each product. Write your answer.
Show your work.

19.
$$\begin{array}{r} 34 \\ \times\ 64 \\ \hline \end{array}$$

21.
$$\begin{array}{r} 101 \\ \times\ 61 \\ \hline \end{array}$$

20.
$$\begin{array}{r} 412 \\ \times\ 35 \\ \hline \end{array}$$

22.
$$\begin{array}{r} 3605 \\ \times\ 52 \\ \hline \end{array}$$

Solve each problem. Show your work. Write your answer.

23. Lori's school district has 3,216 students who travel to school by
bus. If each bus can carry 40 students, how many full buses are
needed to transport these students to school? How many
students will be left?

_____ buses with _____ students left

24. A grocery store has 387 units of canned food. If each shelf can hold
65 units of canned food, how many shelves will be full? How many
units of canned food will be left?

_____ shelves with _____ units of canned food left

Name _____ Date _____

Unit Test

Use each grid to determine how many fractional parts are needed to equal the whole. Write your answer

1.

_____ of the grid is shaded.

_____ parts are needed to equal the whole.

_____ is equal to the whole.

3.

_____ of the grid is shaded.

_____ parts are needed to equal the whole.

_____ is equal to the whole.

2.

_____ of the grid is shaded.

_____ parts are needed to equal the whole.

_____ is equal to the whole.

4.

_____ of the grid is shaded.

_____ parts are needed to equal the whole.

_____ is equal to the whole.

Do the fraction and the decimal match? Write *yes* or *no*.

5. $\frac{1}{5}$ and 0.2 _____

7. $\frac{4}{6}$ and 0.72 _____

6. $\frac{2}{3}$ and 0.6666... _____

8. $\frac{4}{10}$ and 0.41 _____

Write the fractions that are located at the same points on a number line as the following decimals.

9. 0.9 _____

11. −0.2 _____

10. 0.75 _____

12. −2.1 _____

Name _____ Date _____

Unit Test

Write the following numbers as the product of their prime factors. Use Counters, factor trees, mental math, or paper and pencil to determine the prime factors for each number.

13. 55 _____

14. 12 _____

15. 75 _____

Write the following numbers as the product of their prime factors in expanded form, and as an exponential expression.

16. 90 = _____ = _____

17. 40 = _____ = _____

18. 36 = _____ = _____

Use the traditional method to find each product. Write your answer. Show your work.

19. $\begin{array}{r} 37 \\ \times\ 41 \\ \hline \end{array}$

20. $\begin{array}{r} 58 \\ \times\ 43 \\ \hline \end{array}$

Rewrite the fractions in each pair so they have the same denominator, and write a true statement using $<$, $>$, or $=$.

21. $\frac{4}{3}$ $\frac{8}{5}$ _____ _____ _____

22. $\frac{3}{8}$ $\frac{2}{9}$ _____ _____ _____

23. $\frac{2}{5}$ $\frac{5}{7}$ _____ _____ _____

24. $\frac{1}{5}$ $\frac{1}{4}$ _____ _____ _____

25. $\frac{4}{6}$ $\frac{2}{3}$ _____ _____ _____

26. $\frac{2}{5}$ $\frac{10}{25}$ _____ _____ _____

Name _____ Date _____

Unit Test

Add or subtract the rational numbers. Simplify all fractions.

1. $\frac{4}{6} + -\frac{3}{7} =$ _____

2. $0.943 + 0.398 =$ _____

3. $-953 + 939 =$ _____

4. $\frac{7}{4} - \frac{1}{6} =$ _____

5. $-67 - 82 =$ _____

6. $-7.1 + 1.42 =$ _____

Solve the equations.

7. $12^2 =$ _____

8. $5^4 =$ _____

9. $25^1 =$ _____

10. $35^0 =$ _____

11. $9^3 =$ _____

12. $18^2 =$ _____

13. $1^5 =$ _____

14. $10^5 =$ _____

Multiply or divide the rational numbers. Simplify all fractions.

15. $\frac{7}{14} \times -\frac{14}{28} =$ _____

16. $7929 \div 9 =$ _____

17. $2.84 \div -0.4 =$ _____

18. $-8.9 \times -0.44 =$ _____

19. $824 \times 62 =$ _____

20. $\frac{14}{6} \div \frac{6}{7} =$ _____

Name _____ Date _____

Unit Test

Complete the tables below so the percent, fraction, and decimal in each row equal one another. Simplify all fractions.

	Percent	Fraction	Decimal
21.	55%		
22.		$\dfrac{3}{100}$	
23.	50%		0.5
24.		$\dfrac{5}{4}$	

Write the exponential expressions as expanded expressions. Then simplify them by writing them in standard form.

25. 7^{-4} In expanded form, this is _____.

Simplified into standard form, this is _____.

26. 5^{-1} In expanded form, this is _____.

Simplified into standard form, this is _____.

27. 8^{-3} In expanded form, this is _____.

Simplified into standard form, this is _____.

28. 5^{-2} In expanded form, this is _____.

Simplified into standard form, this is _____.

29. 34^{-1} In expanded form, this is _____.

Simplified into standard form, this is _____.

Multiply or divide the exponents. Write the result as an exponential expression.

30. $9^2 \times 9 =$ _____

31. $18^9 \div 18^7 =$ _____

32. $5^4 \times 5^4 =$ _____

33. $7^3 \times 7^{-6} =$ _____

34. $4^4 \div 4^5 =$ _____

35. $6^6 \times 6^5 =$ _____

36. $7^9 \div 7^2 =$ _____

37. $9^8 \div 9^8 =$ _____

Unit Test

Multiply or divide the rational numbers or exponential terms. Simplify all fractions.

1. $-\frac{9}{5} \times \frac{3}{9} =$ _____

2. $391 \div 17 =$ _____

3. $-3.9 \div -1.3 =$ _____

4. $4 \times -79.58 =$ _____

5. $10^3 \times 10^{-6} =$ _____

6. $17^8 \div 17^4 =$ _____

Evaluate each expression by completing the operations inside the parentheses first. Circle the correct answer.

7. $7 + (8 \times 4) =$ 39   60

8. $(1 + 5) \times 2 =$ 11   12

Draw parentheses around the operation that must be performed first in order to get the correct answer.

9. $15 \times 7 - 2 = 75$

10. $12 - 8 \div 2 = 2$

Evaluate the following expressions. Write the solution.

11. $9 \times (7 + 4) + 6^3 =$ _____

12. $4^2 - 8 \times (9 + \frac{1}{2}) =$ _____

Write the solutions to the expressions that are described. It might be useful to use Counters to model the expressions.

13. If you subtract 22 from a number, you get 9 times 3. What is the

number?   _____

14. The home basketball team had 14 less than 2 times the score of the visiting team. If the home team had 68 points, then what score did the visiting team have?   _____ points

Write the solution for the variable. It might be useful to use Counters to model the situations.

15. $-6 + a = 40$ _____

16. $3b = 12$ _____

Complete the table by writing the missing input or output for the expression $0.2x$.

	Input	Output
17.	$x = -20$	
18.		65

Create a table to represent some solutions for the following two-variable equations.

19. $y = x - 9$

x	y

20. $y = 4x$

x	y

Write the number that makes the statement true.

21. $x = v; x + 3 = v +$ _____

22. $u = n; u - 4 = n +$ _____

23. $s = c; 15s =$ _____ c

24. $d = b; 40d = ($ _____ $+ 12)b$

Isolate the variable. Write the solution.

25. $x + 9 = 65$ _____

26. $2t = -92$ _____

Apply the properties to simplify the expressions. Write the resulting expressions.

27. $(9 + 5) + 4$; associative property of multiplication _____

28. $512 + 0$; identity property of addition _____

29. $\frac{1}{3} \times 4 \times 3$; commutative property of multiplication _____

30. $45 + -45$; inverse property of addition _____

Write the solutions to the two-step equations. Verify your solutions by substituting for x. Show your work.

31. $x \div 4 + 10 = 15$ _____

32. $11x - 2 = 31$ _____

33. $-x + 40 = 13$ _____

34. $-9x - 3 = 87$ _____

Name _____ Date _____

Unit Test

Write the solutions to the two-step equations. Verify your solutions by substituting for *x*. Show your work.

1. $\left(\frac{1}{4}\right)x + 8 = 15$

2. $4 - 2x = 40$

Describe how to get from the first point listed to the second point listed. Use a coordinate grid to help you answer the questions. Be sure to describe left-right movement first, and then up-down movement.

3. (9, 4) to (3, 5)

4. (−4, 3) to (−4, 5)

5. (−7, 4) to (6, −7)

6. (2, 6) to (−5, 9)

_____   _____   _____   _____

Plot the given points on the coordinate grids.

7. (−1, 5)

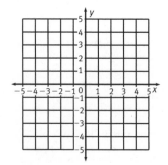

8. (−4, −4)

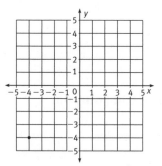

Follow the instructions to create shapes, and then identify the shapes you created.

9. Graph points *A* (−2, 5); *B* (2, 5); and *C* (0, 0). Then connect the points.

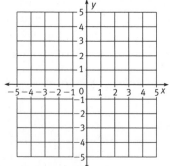

The shape is a(n) _____ .

10. Graph points *A*(−4, 4); *B*(−4, 1); *C*(−4, −2); *D*(2, 4); *E*(2, 1); and *F* (2, −2). Connect points *A*, *B* and *C*, points *D*, *E* and *F*, and points *B* and *E*.

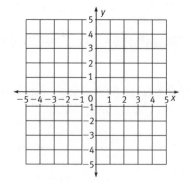

The shape is a(n) _____ .

Determine the length of the line segments that connect these pairs of coordinates. Show your work.

11. (2, 3) and (−2, 3)

12. (0, 6) and (0, 1)

Use the Pythagorean theorem to find the length of each hypotenuse. Show your work below.

13. $a = 5$, $b = 12$, $c =$ _____

14. $a = 11$, $b = 60$, $c =$ _____

Consider these groups of three numbers, and determine whether they are the sides in a right triangle. Explain your answer. You will use the Pythagorean theorem. In each case, the hypotenuse will be identified by c.

15. 10; 24; $c = 26$

16. 2; 4; $c = 5$

_____ _____

Use the formula for the slope given two points, and find the slope of the line.

17. (4, 2) and (2, 6) _____

19. (4, 0) and (6, −8) _____

18. (−2, 2) and (3, 6) _____

20. (0, 6) and (−3, 9) _____

Consider the equation $y = 0.5x + 1$. Answer the following questions about the line that corresponds to this equation.

21. What is the y-intercept of this line? _____

22. Write the coordinates of the y-intercept. _____

23. What is the slope of this line? _____

24. From the y-intercept the slope tells you to move up _____ units,

and then to the _____ 1 unit to find another point on this line.

The coordinates of this point are _____.

Graph the line that corresponds to the equations.

1. $y = 3x + 1$.

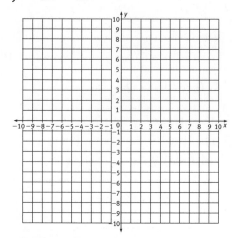

2. $y = x - 4$.

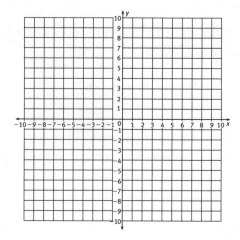

Complete this chart.

Number	Opposite	Reciprocal
3. -9		
4. -1.9		

Simplify each variable expression.

5. $\sqrt[4]{81x^{16}} =$ _____

6. $\sqrt[9]{(1t^{18})} =$ _____

Find the missing leg or hypotenuse. If your answers are not perfect squares, leave them in square-root form. The hypotenuse equals c.

7. $a = 5$; $c = 8$; find b.

$b =$ _____

8. $a = \sqrt{3}$; $b = \sqrt{4}$; find c.

$c =$ _____

9. $b = 4$; $c = 6$; find a.

$a =$ _____

10. $b = \sqrt{25}$; $c = \sqrt{169}$; find a.

$a =$ _____

11. $a = \sqrt{4}$; $c = \sqrt{10}$; find b.

$b =$ _____

12. $a = 3$; $b = 8$; find c.

$c =$ _____

Simplify. Assume all variables do not equal 0.

13. $(15y^2)^{-1} =$ _____

14. $(3^8 7^2)^{\frac{1}{2}} =$ _____

Unit Test

Name _____ Date _____

Simplify using the 5 Basic Rules, PEMDAS, the distributive property, and the rules for signed numbers. Assume all variables do not equal to 0.

15. $\dfrac{9x^4}{-9x^2} =$ _____

16. $(-6x)(36x) =$ _____

17. $(-5)^{-2} \times (-5)^6 =$ _____

18. $\left(\dfrac{45x}{2}\right)^0 =$ _____

19. $(-3x)^4 =$ _____

20. $-z^2(-10z - 1) =$ _____

21. $\dfrac{4x^2}{x^7} =$ _____

22. $-1^{10} =$ _____

Write each step, and be prepared to explain the property, rule, or process that you used.

23. $a + 13 = 20$

24. $4 = k + 31$

25. $-28n - 2 = 138$

26. $-11a - 12 = -1$

27. $6x + 5x - x = 70$

28. $-3y + 2 = 15$

Solve each equation, and prove your answer. Show your work.

29. $5 - 2 = 3a + 7 - a + 4$

$a =$ _____

Check statement: _____

30. $3(a + 2) = 12^2 - 15$

$a =$ _____

Check statement: _____

Solve each inequality, and graph the solution. Watch out for negative coefficients.

31. $-(y - 9) + 2 < 4$

The solution is _____.

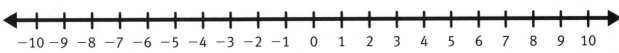

32. $5(y + 2) < 25$

The solution is _____.

Name _____ Date _____

Multiple-Choice Unit Test

Circle the letter of the correct answer.

1. What is the place value where the numbers differ?

5,756   5,776

- **A** Ones
- **B** Tens
- **C** Hundreds
- **D** Thousands

2. What is the place value where the numbers differ?

8,410   8,418

- **A** Ones
- **B** Tens
- **C** Hundreds
- **D** Thousands

3. 7363
    + 427

- **A** 7,790
- **B** 7,770
- **C** 7,780
- **D** 7,870

4. 8826
    + 976

- **A** 8,702
- **B** 8,802
- **C** 9,792
- **D** 9,802

5. 1122
    + 49

- **A** 1,171
- **B** 1,181
- **C** 1,161
- **D** 1,155

6. 3421
    − 987

- **A** 2,434
- **B** 2,435
- **C** 2,444
- **D** 2,534

7. 2744
    − 845

- **A** 2001
- **B** 1909
- **C** 1999
- **D** 1899

8. What addition problem could you use to check the following subtraction problem?

4933
− 55
4878

- **A** 4878
     + 55
     4933

- **B** 4878
     − 55
     4933

- **C** 4933
     + 4878
     55

- **D** 4933
     + 55
     4878

9. Which of the following is equivalent to the product of 22 × 97?

   A 22 × (90 + 7)
   B 22 × (97 + 7)
   C (22 − 2) × 97
   D All of the above

10. Which of the following is equivalent to the product of 67 × 8?

   A 8 × (67 + 1)
   B 8 × (60 + 7)
   C (8 + 10) × 57
   D All of the above

11. Which of the following is equivalent to the product of 31 × 19?

   A 31 × (10 + 9)
   B 31 × (12 + 7)
   C (30 + 1) × 19
   D All of the above

12. 113
      × 37

   A 4,181
   B 150
   C 1,130
   D 3,054

13. 421
      × 12

   A 5,052
   B 433
   C 1,263
   D 4,508

14. 69
      × 28

   A 690
   B 1,932
   C 97
   D 2,464

15. 54648 ÷ 9

   A 491,832
   B 6,072
   C 0.00016469
   D 54639

16. A grocery store has 233 units of canned food. If each shelf can holds 52 units of canned food, how many units of canned food will be left after you stock only full shelves?

   A 61
   B 1
   C 25
   D 41

Name _____ Date _____

Multiple-Choice Unit Test

Circle the letter of the correct answer.

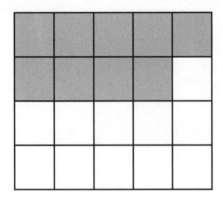

1. What fraction of the grid is shaded?

A $\frac{11}{19}$

B $\frac{20}{11}$

C $\frac{9}{20}$

D $\frac{11}{20}$

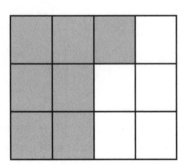

2. What fraction of the grid is shaded?

A $\frac{5}{12}$

B $\frac{12}{7}$

C $\frac{7}{12}$

D $\frac{5}{7}$

3. Match the decimal 1.2 to a fraction.

A $\frac{12}{1}$

B $\frac{1}{12}$

C $\frac{6}{5}$

D $\frac{5}{6}$

4. Match the decimal 0.15 to a fraction.

A $\frac{1}{5}$

B $\frac{15}{1}$

C $\frac{0}{15}$

D $\frac{3}{20}$

5. Match the fraction $\frac{1}{8}$ to a decimal.

A 0.0125

B 0.125

C 0.18

D 0.81

6. Match the fraction $\frac{2}{3}$ to a decimal.

A 0.3232...

B 0.2323...

C 0.6666...

D All of the above

7. Write 50 as a product of prime factors using expanded notation.

A $2 \times 5 \times 5$

B 2×25

C 5^0

D 2×5

8. Write 72 as a product of prime factors using expanded notation.

A 2×3

B $2 \times 2 \times 2 \times 3 \times 3$

C 72

D $2^2 \times 3^2$

9. Write 20 as a product of prime factors using exponents.

A $2^2 \times 5$

B 4×5

C 2^0

D 2×5

10. Write 52 as a product of prime factors using exponents.

A $2^2 \times 13$

B 2×26

C 5^2

D 2×13

11. $\quad 422$
$\quad \underline{\times\ 14}$

A $2{,}110$

B $6{,}912$

C 436

D $5{,}908$

12. $2088 \div 8$

A 0.0038

B 21

C 261

D 2000

13. Rewrite the fractions $\frac{5}{3}$ and $\frac{8}{5}$ so they have the same denominator.

A $\frac{5}{15}$ and $\frac{3}{15}$

B $\frac{25}{15}$ and $\frac{24}{15}$

C $\frac{13}{8}$ and $\frac{8}{8}$

D $\frac{5}{3}$ and $\frac{8}{5}$

14. What symbol should be placed in the blank between the two fractions to make the relationship true?

$\frac{8}{24} \ \underline{\qquad} \ \frac{2}{3}$

A $<$

B $>$

C $=$

D None of the above

Name _____ Date _____

Multiple-Choice Unit Test

Circle the letter of the correct answer.

1. $\frac{2}{3} + \frac{6}{8}$

    A $\frac{8}{8}$ or 1

    B $\frac{17}{12}$

    C $\frac{8}{11}$

    D $2\frac{6}{8}$

2. $9.26 - 9.38$

    A 18.64

    B 0.12

    C -0.12

    D 4.88

3. $\frac{2}{9} - \frac{7}{2}$

    A $\frac{17}{9}$

    B $-\frac{5}{9}$

    C $-\frac{5}{7}$

    D $-\frac{59}{18}$

4. 20^2 is equivalent to which of the following numbers?

    A 1

    B 40

    C 202

    D 400

5. 8^3 is equivalent to which of the following numbers?

    A 83

    B 24

    C 1

    D 512

6. 15^0 is equivalent to which of the following numbers?

    A 0

    B 1

    C 15

    D 150

7. 8.69×2.7

    A 1.39

    B 25.3

    C 23.463

    D 234.63

8. 4.85×0.04

    A 0.0194

    B 0.194

    C 1.194

    D 11.94

9. Find the quotient.

$\frac{2}{3} \div \frac{3}{7}$

A $\frac{2}{7}$

B $\frac{7}{2}$

C $\frac{9}{14}$

D $\frac{14}{9}$

10. Match 45% to an equivalent fraction.

A $\frac{45}{10}$

B $\frac{5}{4}$

C $\frac{4}{5}$

D $\frac{9}{20}$

11. Match 23% to an equivalent decimal

A 2.3

B 0.23

C 0.2323...

D 23.0

12. Match $\frac{5}{4}$ to an equivalent decimal

A 5.4

B 0.8

C 1.1

D 1.25

13. Which of the following is the expanded form of 4^{-3}?

A $\frac{1}{4} \times \frac{1}{4} \times \frac{1}{4}$

B $\frac{1}{4^3}$

C 0.444

D -43

14. Which of the following is the expanded form of 8^{-5}?

A 0.88888

B $\frac{1}{8} \times \frac{1}{8} \times \frac{1}{8} \times \frac{1}{8} \times \frac{1}{8}$

C $\frac{1}{8^5}$

D -40

15. $5^9 \times 5^2$

A 500

B 5^6

C 5^{16}

D 5^{11}

16. $12^3 \div 12^4$

A 12^1

B 12^{-1}

C 12^7

D $12^{\frac{3}{4}}$

Name _____ Date _____

Multiple-Choice Unit Test

Circle the letter of the correct answer.

1. 2.36×0.431

   A 0.101716
   B 1.01716
   C 10.1716
   D 101.716

2. $\frac{2}{9} \div \frac{1}{8}$

   A 36

   B $\frac{1}{36}$

   C $\frac{16}{9}$

   D $\frac{9}{16}$

3. Evaluate the expression by completing the operations in the correct order.

   $15 - (2 - 4)$

   A 7
   B 9
   C 13
   D 17

4. What should the parentheses be placed around to get the correct answer?

   $6 \times 5 + 1 = 36$

   A $5 + 1$
   B 6×5
   C $+$
   D $\times$

5. $7 \times (2 + 8) + 3^4$ can be simplified to which of the following?

   A 25
   B 103
   C 151
   D 390,625

6. If you subtract 13 from a number, you get 24 divided by 2. What is the number?

   A 25

   B $-\frac{11}{2}$

   C $\frac{11}{2}$

   D 33

7. Solve for the variable.

   $3a = 1662$

   A 4986
   B 1665
   C 1659
   D 554

8. What is the input for the expression $x \times \frac{1}{3}$ if the output is 10?

   A -4
   B 4
   C 15
   D 30

Name _____ Date _____

Multiple-Choice Unit Test, continued

X	Y
0	4
1	−3
2	−10

9. This table is a possible solutions set for which of the following equations?

A $y = -3x + 4$
B $y = 3x + 4$
C $y = -7x + 4$
D $y = 7x + 4$

10. Which numbers makes the following equations true?

$x = v; x \times \frac{1}{4} = v \div$ _____

A 8

B 16

C $\frac{1}{4}$

D 4

11. Solve for the variable.

$x - 6 = -10$

A −16
B −4
C 10
D 16

12. Name the property that is used to change Expression 1 into Expression 2.

Expression 1: $8 \times 12 \times 5$

Expression 2: $8 \times 5 \times 12$

A Identity property of multiplication
B Inverse property of multiplication
C Commutative property of multiplication
D Associative property of multiplication

13. Solve for the variable.

$3x + 6 = 3$

A $\frac{1}{3}$
B −1
C 1
D $\frac{1}{6}$

14. Solve for the variable.

$-7x + 6 = -29$

A 3
B −5
C 5
D −3

Name _____ Date _____

Multiple-Choice Unit Test

Circle the letter of the correct answer.

1. Solve for the variable.

   $-x + 5 = 7$

   A -2

   B 2

   C -12

   D 12

2. $x = 7$ is the solution to which of the following equations?

   A $2x + 3 = 17$

   B $-14x + 4 = 2$

   C $35x + 2 = 7$

   D $12x - 80 = 4$

3. How do you move from $(-5, 1)$ to $(-4, 4)$?

   A Left 2 units and up 3 units

   B Right 2 units and down 3 units

   C Right 1 unit and up 3 units

   D Left 3 units and up 2 units

4. How do you move from $(9, -1)$ to $(6, 5)$?

   A Left 3 units and up 6 units

   B Right 3 units and down 6 units

   C Right 6 unit and up 3 units

   D Left 6 unit and up 3 units

5. What point is graphed on the following coordinate grid?

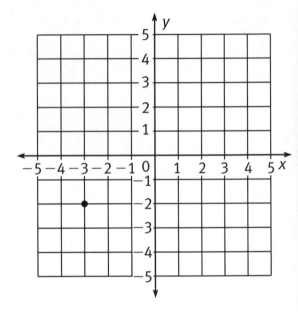

   A $(-2, -3)$

   B $(-3, -2)$

   C $(2, 3)$

   D $(3, 2)$

6. Which of the following points could be connected to create a rectangle?

   A $(9, 10), (3, 5), (9, 5), (3, 10)$

   B $(1, 8), (8, 1), (4, 5), (5, 4)$

   C $(1, 1), (2, 2), (3, 3), (4, 4)$

   D $(4, 2), (2, 0), (0, -2), (-2, -4)$

7. Determine the length of the line segments that connects $(4, 3)$ and $(-4, 3)$

  A   0

  B -8

  C   8

  D   16

8. Determine the length of the line segments that connects $(2, 7)$ and $(2, 1)$

  A   0

  B -6

  C   6

  D   8

9. Find the length of the hypotenuse if the lengths of the other two sides are 5 and 12.

  A   13

  B   14

  C   17

  D   60

10. Find the length of the side if the length of the hypotenuse is 5, and the length of the other side is 4.

  A   3

  B   4

  C   5

  D   9

11. Use the formula for the slope given two points at $(2, 5)$ and $(-2, 8)$

  A $\dfrac{3}{4}$

  B $-\dfrac{3}{4}$

  C $\dfrac{4}{3}$

  D $-\dfrac{4}{3}$

12. Which of the following is a graph of $y = -\dfrac{1}{5}x + 3$?

  A

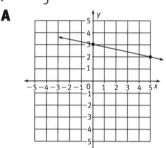

  B

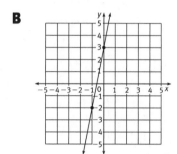

  C

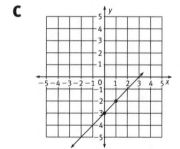

  D

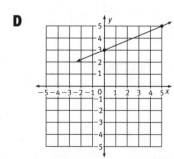

Name _____ Date _____

Multiple-Choice Unit Test

Circle the letter of the correct answer.

1. Which of the following is a graph of $y = 2x - 1$?

A

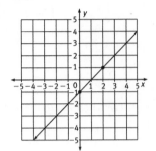

B

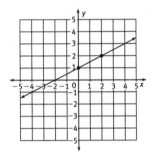

C

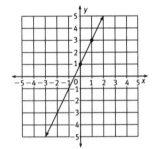

D

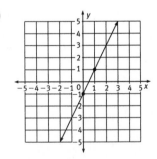

2. Match 12 with its opposite.

A $-\dfrac{1}{12}$

B $\dfrac{1}{12}$

C -12

D 12

3. Match -4 with its reciprocal.

A $\dfrac{1}{4}$

B -4

C 4

D $-\dfrac{1}{4}$

4. Simplify the variable expression.

$$\sqrt{9x^6y^2}$$

A $9xy$

B $3x^3y$

C $3xy$

D $3x^4y^0$

5. Simplify the variable expression.

$$\sqrt{10^2x^2}$$

A $100x^{-1}$

B $10x$

C $10x^2$

D $100x^2$

6. Find the missing side of the right triangle. The hypotenuse is c.

$a = 7$, $c = 16$, find b.

A 9

B 14

C $\sqrt{9}$

D $\sqrt{207}$

7. Find the missing side of the right triangle. The hypotenuse is c.

$a = 12$, $b = 13$, find c.

A -5
B $\sqrt{313}$
C 5
D $\sqrt{5}$

8. Simplify. Assume all variables do not equal zero.

$4x^2 + 7x^3 + 2x^2 - x^3$

A $6x^2 + 6x^3$
B $12x^5$
C $12x^6$
D Can not be further simplified

9. Simplify. Assume all variables do not equal zero.

$3(x + 7) + 2x$

A $5x + 21$
B $5x + 7$
C $23x$
D Can not be further simplified

10. Simplify. Assume all variables do not equal zero.

$x + x^2 + 3x^3$

A $5x^6$
B $5x^7$
C $x + 3x^6$
D Can not be further simplified

11. Simplify. Assume all variables do not equal zero.

$(24x^2)^3$

A $24x^5$
B 24^3x^5
C 24^3x^6
D $24x^3$

12. Solve for the variable.

$5x - 2x + 3x = 7x - 10$

A 10
B -10
C 20
D -20

13. Solve for the variable.

$5(x + 7) = 35$

A $\frac{5}{7}$
B -7
C 0
D 7

14. Solve for the inequality.

$2y - 7y + 3y < 10$

A $y < -5$
B $y < 5$
C $y = 5$
D $y > -5$

Multiple-Choice Unit Test Answers

Unit 1 pages 128–129

1.	B	2.	A	3.	A	4.	D
5.	A	6.	A	7.	D	8.	A
9.	A	10.	B	11.	D	12.	A
13.	A	14.	B	15.	B	16.	C

Unit 2 pages 130–131

1.	C	2.	C	3.	C	4.	D
5.	B	6.	C	7.	A	8.	B
9.	A	10.	A	11.	D	12.	C
13.	B	14.	A				

Unit 3 pages 132–133

1.	B	2.	C	3.	D	4.	D
5.	D	6.	B	7.	C	8.	B
9.	D	10.	D	11.	B	12.	D
13.	A	14.	B	15.	D	16.	B

Unit 4 pages 134–135

1.	B	2.	C	3.	D	4.	A
5.	C	6.	A	7.	D	8.	D
9.	C	10.	D	11.	B	12.	C
13.	B	14.	C				

Unit 5 pages 136–137

1.	A	2.	D	3.	C	4.	A
5.	B	6.	A	7.	C	8.	C
9.	A	10.	A	11.	B	12.	A

Unit 6 pages 138–139

1.	D	2.	C	3.	D	4.	B
5.	B	6.	D	7.	B	8.	A
9.	A	10.	D	11.	C	12.	A
13.	C	14.	D				

If students do not show understanding of at least 75% of the items in Unit 1, you may choose to review Unit 1 and reassess before continuing to Unit 2.

If students do not show understanding of at least 75% of the items in Unit 2, you may choose to review Unit 2 and reassess before continuing to Unit 3.

If students do not show understanding of at least 75% of the items in Unit 3, you may choose to review Unit 3 and reassess before continuing to Unit 4.

If students do not show understanding of at least 75% of the items in Unit 4, you may choose to review Unit 4 and reassess before continuing to Unit 5.

If students do not show understanding of at least 75% of the items in Unit 5, you may choose to review Unit 5 and reassess before continuing to Unit 6.

If students do not show understanding of at least 75% of the items in Unit 6, you may choose to review Unit 6 before recommending entry into an Algebra Program.

Placement Test Answers

Use the answers below to assess students' knowledge of *Algebra Readiness* content.

If students do not successfully complete 75% of the items on the test the student should be placed into the *Algebra Readiness* program. If students successfully complete 75% of the items, you may choose to have them begin in the *Algebra Readiness* program or place them into Algebra I.

Unit 1   page 14

❶ C   **❷** A   **❸** D   **❹** A   **❺** D   **❻** B   **❼** D   **❽** C

Unit 2   page 15

❶ B   **❷** D   **❸** A   **❹** D   **❺** C   **❻** B

Unit 3   page 16

❶ D   **❷** C   **❸** B   **❹** B   **❺** B   **❻** C   **❼** A   **❽** C

Unit 4   page 17

❶ B   **❷** D   **❸** C   **❹** C   **❺** B   **❻** B   **❼** C   **❽** C

Unit 5   page 18

❶ A   **❷** A   **❸** B   **❹** C   **❺** A   **❻** D

Unit 6   page 19

❶ B   **❷** A   **❸** D   **❹** B   **❺** B   **❻** C   **❼** D   **❽** A

Unit Test Answers

Unit 1 Pages 116–117

1. hundreds
2. tens
3. ones
4. thousands
5. 5,769
6. 3,098
7. 1,470
8. 1,316
9. 2,169
10. 854
11. 601
12. 1894
13. 1,960
14. 2,692
15. 20; 8; 224
16. 50; 2; 624
17. 20; 7; 108
18. 60; 7; 201
19. $\begin{array}{r} 136 \\ +2040 \\ \hline 2,176 \end{array}$
20. $\begin{array}{r} 2060 \\ +12,360 \\ \hline 14,420 \end{array}$
21. $\begin{array}{r} 101 \\ +6060 \\ \hline 6,161 \end{array}$
22. $\begin{array}{r} 7210 \\ +180250 \\ \hline 187,460 \end{array}$
23. 80; 16
24. 5; 62

Unit 2 Pages 118–119

1. $\frac{7}{24}, \frac{17}{24}, \frac{24}{24}$
2. $\frac{12}{15}, \frac{3}{15}, \frac{15}{15}$
3. $\frac{9}{10}, \frac{1}{10}, \frac{10}{10}$
4. $\frac{3}{8}, \frac{5}{8}, \frac{8}{8}$
5. yes
7. no
6. yes
8. no
9. $\frac{9}{10}$
10. $\frac{3}{4}$
11. $-\frac{1}{5}$
12. $-\frac{21}{10}$
13. $55 = 5 \times 11$
14. $12 = 2 \times 2 \times 3$
15. $75 = 3 \times 5 \times 5$
16. $2 \times 3 \times 3 \times 5; 2 \times 3^2 \times 5$
17. $2 \times 2 \times 2 \times 5; 2^3 \times 5$
18. $2 \times 2 \times 3 \times 3; 2^2 \times 3^2$
19. $\begin{array}{r} 37 \\ +1480 \\ \hline 1,517 \end{array}$
20. $\begin{array}{r} 174 \\ +2320 \\ \hline 2,494 \end{array}$
21. $\frac{20}{15}; <; \frac{24}{15}$
22. $\frac{27}{72}; >; \frac{16}{72}$
23. $\frac{14}{35}; <; \frac{25}{35}$
24. $\frac{4}{20}; <; \frac{5}{20}$
25. $\frac{12}{18}; =; \frac{12}{18}$
26. $\frac{50}{125}; =; \frac{50}{125}$

Unit 3 Pages 120–121

1. $\frac{5}{21}$
2. 1.341
3. -14
4. $1\frac{7}{12}$
5. -149
6. -5.68
7. 144
8. 625
9. 25
10. 1
11. 729
12. 324
13. 1
14. 100,000
15. $-\frac{1}{4}$
16. 881
17. -7.1
18. 3.916
19. 51,088
20. $\frac{49}{18}$
21. $\frac{11}{20}; 0.55$
22. 3%; 0.03
23. 50%; $\frac{1}{2}$
24. 125%; 1.25
25. $\frac{1}{7} \times \frac{1}{7} \times \frac{1}{7} \times \frac{1}{7}; \frac{1}{2401}$
26. $\frac{1}{5}; \frac{1}{5}$
27. $\frac{1}{8} \times \frac{1}{8} \times \frac{1}{8}; \frac{1}{512}$
28. $\frac{1}{5} \times \frac{1}{5}; \frac{1}{25}$
29. $\frac{1}{34}; \frac{1}{34}$
30. 9^3
34. 4^{-1}
31. 18^2
35. 6^{11}
32. 5^8
36. 7^7
33. 7^{-3}
37. 9^0; or 1

Unit Test Answers, continued

Unit 4 Pages 122–123

1. $-\dfrac{3}{5}$ **2.** 23

3. 3 **4.** −318.32

5. 10^{-3} **6.** 17^4

7. 39 **8.** 12

9. $(7 - 2)$ **10.** $(12 - 8)$

11. 315 **12.** −60

13. 49 **14.** 41

15. $a = 46$ **16.** $b = 4$

17. −4 **18.** $x = 325$

19.

x	0	1	2	3
y	−9	−8	−7	−6

20.

x	0	1	2	3
y	0	4	8	12

21. 3 **22.** −4

23. 15 **24.** 28

25. $x = 56$ **26.** $t = -46$

27. $9 + (5 + 4)$

28. 512

29. $\dfrac{1}{3} \times 3 \times 4$

30. 0

31. $x = 20$ **32.** $x = 3$

33. $x = 27$ **34.** $x = -10$

Unit 5 Pages 124–125

1. $x = 28$

2. $x = -18$

3. left 6 units, and up 1 unit

4. no movement left or right, and up 2 units

5. right 13 units, and down 11 units

6. left 7 units, and up 3 units

7. set a point at $(-1, 5)$

8. set a point at $(-4, -4)$

9. Connect points A $(-2, 5)$, B $(2, 5)$, and C $(0, 0)$; triangle

10. Connect points A $(-4, 4)$, B $(-4, 1)$, C $(-4, -2)$ D $(2, 4)$, E $(2, 1)$, F $(2, -2)$; the letter H

11. $2 - (-2) = 4$

12. $6 - 1 = 5$

13. 13 **14.** 61

15. yes; because $10^2 + 24^2 = 26^2$; $100 + 576 = 676$

16. 2; 4; $c = 5$ no; because $2^2 + 4^2 \neq 5^2$; $4 + 16 \neq 25$

17. −2

18. $\dfrac{4}{5}$ **21.** 1

19. −4 **22.** $(0, 1)$

20. −1 **23.** 0.5, or $\dfrac{1}{2}$

24. 0.5; right; $(1, 1.5)$

Unit 6 Pages 126–127

1. Students should set a line going through $(0, 1)$ and $(-3, 8)$.

2. Students should set a line going through $(0, -4)$ and $(2, -2)$.

3. $9, -\dfrac{1}{9}$

4. $1.9, -\dfrac{10}{19}$

5. $3x^4$ **6.** t^2

7. $\sqrt{39}$ **8.** $\sqrt{7}$

9. $\sqrt{20}$ **10.** 12

11. $\sqrt{6}$ **12.** $\sqrt{73}$

13. $\dfrac{1}{15y^2}$ **14.** $3^4 7 = 567$

15. $-x^2$ **16.** $-6^3 x^2$

17. -5^4 **18.** 1

19. $81x^4$ **20.** $10z^3 + z^2$

21. $\dfrac{4}{x^5}$ **22.** −1

23. $a = 7$ **24.** $k = -27$

25. $n = -5$ **26.** $a = -1$

27. $x = 7$ **28.** $-\dfrac{13}{3}$

29. $-4, 3 = 3$

30. $41, 129 = 129$

31. $y > 7$, set open circle on 7 with arrow to the right

32. $y < 3$, set open circle on 3 with arrow to the left

Student Assessment Record

		Unit 1				Unit 2				Unit 3				Unit 4				Unit 5				Unit 6			
		1	2	3	4	5	6	7	8	9	10	11	12	13	14	15	16	17	18	19	20	21	22	23	24
Student Book Completed																									
Informal Assessment **Computing**																									
Informal Assessment **Understanding**																									
Informal Assessment **Applying**																									
Informal Assessment **Reasoning**																									
Informal Assessment **Engaging**																									
Informal Assessment **Problem Solving**																									
Formative Assessment **Diagnositic Test**	Lesson 1A	3	3	3	3	3	3	3	3	3	3	3	3	3	3	3	3	3	3	3	3	3	3	3	3
	2A	3	3	3	3	3	3	3	3	3	3	3	3	3	3	3	3	3	3	3	3	3	3	3	3
	3A	3	3	3	3	3	3	3	3	3	3	3	3	3	3	3	3	3	3	3	3	3	3	3	3
	4A	3	3	3	3	3	3	3	3	3	3	3	3	3	3	3	3	3	3	3	3	3	3	3	3
Formal Assessment **Chapter Test**	Lesson 1																								
	2																								
	3																								
	4																								
Formal Assessment **Unit Test**																									